Acclaimed textile artist Ann Goddard takes three-dimensional textiles to a new level with mixed-media work.

In this practical book she combines textile and non-textile elements with construction. She talks the reader through contrasting materials such as linen, loose fibres, paper and yarn with concrete and wood. Fragile is juxtaposed with hard, natural with man-made, and beauty with imperfection. Techniques range from stitching, wrapping, couching and knotting to sawing, drilling and casting. Beautifully illustrated throughout, this is an essential book for any embroiderer or textile artist looking for fresh ideas and techniques.

Mixed Media Textile Art in three dimensions

Mixed Media Textile

Art in three dimensions

Ann Goddard

BATSFORD

PREVIOUS PAGE *Habitat* series; one of four constructions; bark fragments, neoprene cord, copper wire (W38 x H56 x D20 cm / W15 x H22 x D8 in)

For Phil, Sam and Sophie

First published in the United Kingdom in 2022 by
B.T. Batsford
43 Great Ormond Street
London WC1N 3HZ

An imprint of B.T. Batsford Holdings Ltd

ISBN: 9781849946926

A CIP catalogue record for this book is available from the British Library.

10 9 8 7 6 5 4 3 2 1

Reproduction by Rival Colour Ltd, UK
Printed by Toppan Leefung Ltd, China

This book can be ordered direct from the publisher at www.batsfordbooks.com, or try your local bookshop.

Contents

Introduction

Creativity involves breaking out of established
patterns in order to look at things in a different way

EDWARD DE BONO

Concrete, wood, lead and bark might seem unlikely materials to be included in a book aimed at textile artists, but these are just a few of the materials that I have incorporated into my textile art over the years. A friend described me as working with 'extreme mixed media'.

Since I started learning how to embroider, the assumptions of what textile practice can be have changed radically. My own creative process has also evolved and grown over the years.

Working with mixed media challenges the traditional perception of textiles as a separate genre. This book will explore three-dimensional reliefs, assemblages and constructions that combine a variety of mixed media with textiles and/or textile techniques. It will give an insight into my own work, including where my inspiration comes from and how I interpret ideas through materials, and will show my approach to developing work from concept to outcome.

Using examples of my work, this book will guide you from theme and inspiration through to materials, tools, techniques, and finally to construction. It will also demonstrate how to add meaning to your art. By sharing my creative process I hope I can encourage you to incorporate unconventional methods into your work.

In addition to my pieces, the book features the work of some inspiring and exciting artists creating three-dimensional constructions. Their practice incorporates all manner of eclectic materials: ocean debris, litter, concrete, kelp, handmade paper, tacks, barbed wire, clay, plastics, metal, beach finds, toy soldiers and repurposed items. All combine textile elements or textile techniques with these materials.

The type of art that I make can be described broadly as follows:

- is three-dimensional
- is constructed
- features a wide range of materials
- uses textiles or textile analogues
- incorporates (often highly) contrasting materials
- features extensive incorporation of repetitive elements
- has strong structural elements
- uses materials to explore ideas

These criteria give me a huge range of possibilities for experimentation in my work, as you will see from the varied pieces included in this book.

I often refer to my own works as constructions rather than sculptures, as people tend to associate the term 'sculpture' with traditional wood carvings, stone carvings or bronze statues. Construction describes exactly what I do: I build up forms and surfaces by putting parts together.

My background

I originally studied art with a focus in ceramics while training to be a teacher, but after college found it difficult to carry on without access to a kiln. It was much later, when I was in my forties, that I discovered textiles as an art form.

Stumbling upon a City & Guilds creative embroidery course was a seminal experience for me which changed my life. I was immediately hooked. It was so exciting and different from the functional sewing I had experienced at school. It was a revelation to me that you could leave frayed edges and loose threads. Experimentation was encouraged, and the tutors were inspirational. It became obvious that there are no rights or wrongs in art, only possibilities, and the only limit to what is possible is our own imagination. Winning a City & Guilds Medal of Excellence motivated me to carry on.

Joining E2 (E Squared), a local textile group based on the Wirral, enabled me to become more involved with textiles and provided an opportunity to exhibit. It was inspiring to be part of a group of friendly, supportive and like-minded people.

A further course in stitched and constructed textiles followed. This encouraged combining textiles and stitch with materials and techniques from the other departments, including wood, glass, print, ceramics and metal. I was so excited by the visual juxtaposition of these materials: it was an epiphany.

Winning a Charles Henry Foyle Trust Award for Stitched and Constructed Textiles allowed me to take an MA in Fine Art, where the emphasis was on working with textiles in a fine art context. This provided me with knowledge of the wider art world, and how to select materials, processes and methods of construction to explore ideas and convey significance and meaning beyond the visual.

My hybrid practice

Using textile tradition as a starting point, my practice has expanded into one that can be described as hybrid. Previously separate art media are combined to create eclectic works; boundaries are crossed, merged and broken, challenging the expectations of what a textile is. Professor Polly Binns has described this type of practice as 'exploding the category'. Textile artist Michael Brennand-Wood regards it as challenging 'the boundaries of textile and craft approaches'.

Categorisation could be regarded as restricting possibilities. Artist Maggie Henton views categorisation as setting 'false boundaries' around her work and limiting 'ways of seeing and interpreting it'. Author Laurie Britton Newell suggests that textiles 'is not a separate category but an ingredient, a process, something that has enabled the making of idea into visible things'.

The further removed my own work becomes from the traditional definition of textiles, the harder

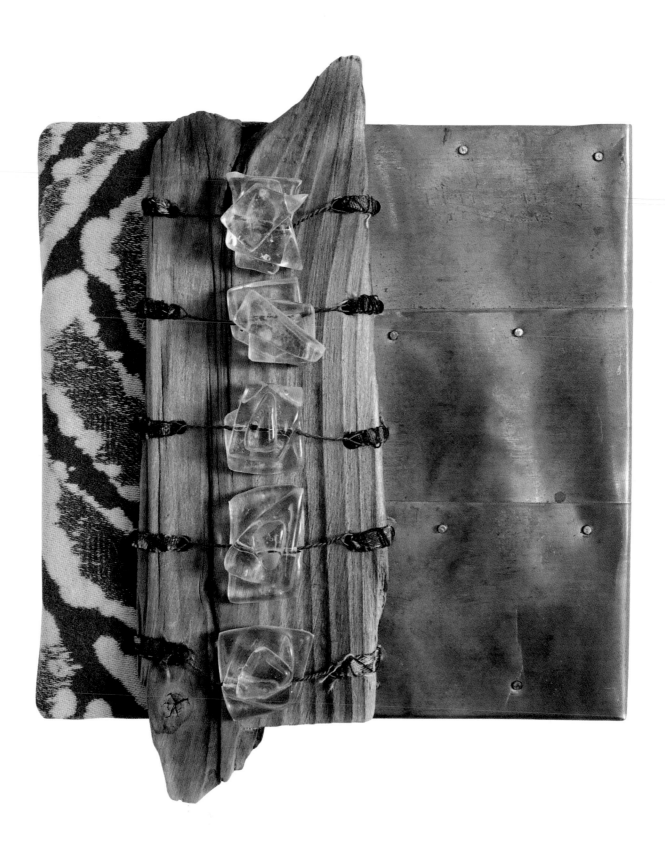

it becomes to contextualise. Initially I found that this limited exhibition opportunities and it was something I often worried about. Now I realise my work exists forever on the margins of textile practice. By manipulating diverse materials, I have developed my own visual language and I am no longer concerned with categorisation – my work is what it is!

I hope the book will appeal to anyone who is interested in alternative approaches to textile practice: those who have textile or embroidery experience and are keen to try something different; those who enjoy being experimental; anyone already working with mixed media and is curious about how others approach it; and those who are fascinated by three-dimensional artwork. I love the visual excitement that comes from juxtaposing different materials and the challenge of finding new ways to join them. I hope I can pass on some of my enthusiasm to you.

LEFT Small construction, an example of one of my first pieces combining mixed media; bleach printed fabric, wood, glass, tin and copper wire on wood base (W14 x H17 x D4 cm / W5½ x H6¾ x D1½ in)

1

Exploring themes

_____

LEFT *Penned In* series (detail): slate, felted silk noil paper, sticks, threaded rods, wing nuts

Exploring themes

An artist is an explorer

HENRI MATISSE

Exploring a theme is an essential starting point for me. A theme provides inspiration and a focus for creativity. My pieces evolve from self-directed themes that are meaningful to me.

Two ongoing self-directed themes that I have explored thoroughly are 'boundaries' and 'human impact on the environment'. Most of my work has evolved from these themes. I keep returning to them to explore the ideas in depth. As the theme grows, ideas tend to be refined. I find that revisiting often inspires different interpretations.

When I explore a theme, the artistic outcomes vary. I might create individual pieces, a piece made up of multiple modular units, or a series (several variations developed from the same aspect of the theme). The constructions developed from each of these ongoing themes are quite different. Those from the 'boundaries' theme are rigid, ordered and controlled; in contrast, those developed from the environmental theme show organic qualities.

The subsequent examples show how I explore a theme and offer suggestions of ways in which you could explore a theme.

Self-directed theme: boundaries

The theme of 'boundaries' came about as I tried to understand where my practice fitted in the art world. My style of working is difficult to categorise as it straddles boundaries between textiles and sculpture, and between fine art and craft.

When I was studying fine art, I became aware of the hierarchies between art and craft. I found it dismaying that textile art, with its connection to women and craft processes, was regarded as less worthy than traditional painting and sculpture. I wanted to understand why this should be.

I questioned the role of a boundary in different contexts, discovering that socially constructed boundaries and physical boundaries in the landscape have the same function: to define difference; to prevent intrusion from outside or straying from within; to restrict and control. I used these qualities to develop the following two works: *Boundary Lines* and *Penned In*.

Boundary Lines

This piece began from research into the concept of socially constructed boundaries.

Idea to use materials and methods of construction to reflect restriction and control
Hard materials planks of wood, slate, metal mesh, willow, nuts, bolts, screws, mending plates
Textile materials scrim, silk noil fibres, neoprene cord in different thicknesses, paper yarn
Techniques sawing wood, cutting slate, handmade silk paper, trapping

Boundary Lines consists of seven wall-hung units; each modular unit is similar but different. It has been exhibited in two different arrangements. In

RIGHT *Boundary Lines*; examples of the different materials and joining techniques

the first, the units were hung with spaces between them, suggesting fencing; in the second, the units were butted against each other, forming a barrier. The materials and processes used are intended to elicit possible associations and connotations. The fixings represent traditional masculinity and have been chosen to hold back the textile elements. The fabric, handmade paper, fibres and thread refer to stereotypes of women's work or craft. Combining skills and materials from different disciplines was a deliberate choice to reject the notion of boundaries.

Penned In series

This series of three separate pieces also developed from research into boundaries. The concept behind them is influenced by the way that boundaries in the landscape are intended to both prevent intrusion from outside and straying from within, as a metaphor for the hierarchies in art.

Idea to use materials and methods of construction to reflect the restrictive and controlling nature of hedges and fences
Hard materials wood, slate, twigs, threaded steel rods, nuts and washers
Textile materials silk noils
Techniques sawing and burning wood, cutting slate, drilling wood and slate, handmade silk paper making

The slate references the slate fences found in North Wales not far from where I live, the twigs signify hedges, and the felted paper alludes to traditional textiles. The contrasting materials have been layered and clamped together to control and restrain the different elements. The juxtaposition of hard materials traditionally worked by men with textile suggests that boundaries are not immutable.

Self-directed theme: human impact on the environment

This theme evolved from a lifelong interest in nature. Like many artists, I have a broad collection of curious items picked up when out walking in woods or on beaches: pieces of wood, bark, stones and fossils that I keep for inspiration.

I happened upon a large quantity of bark when visiting a wood surgeon's market garden. I was told it had come from an ancient yew tree, inadvertently killed during the construction of a housing development. And just like that a 500-year-old tree is lost along with a whole ecosystem relying on its existence. I was drawn to the colour, texture and shape of the bark pieces, and its age made it seem especially precious. I could not let it go to waste, so I recycled it by incorporating it into my artwork.

It was while researching the theme of human impact on the environment that I came across a quote from environmental artist Andy Goldsworthy: 'Our lives and what we do affect Nature so closely that we cannot be separate from it.' This seemed so relevant, especially regarding the reason behind the demise of the yew tree, and led me to develop a body of work investigating local and global ecological issues, including the destruction of habitats, extinction of species, and loss of biodiversity.

LEFT *Penned In* series; slate, wood, twigs, felted silk noil paper, threaded rods and wing nuts (W20 x H100 x D4 cm / W8 x H39¼ x D1½ in)

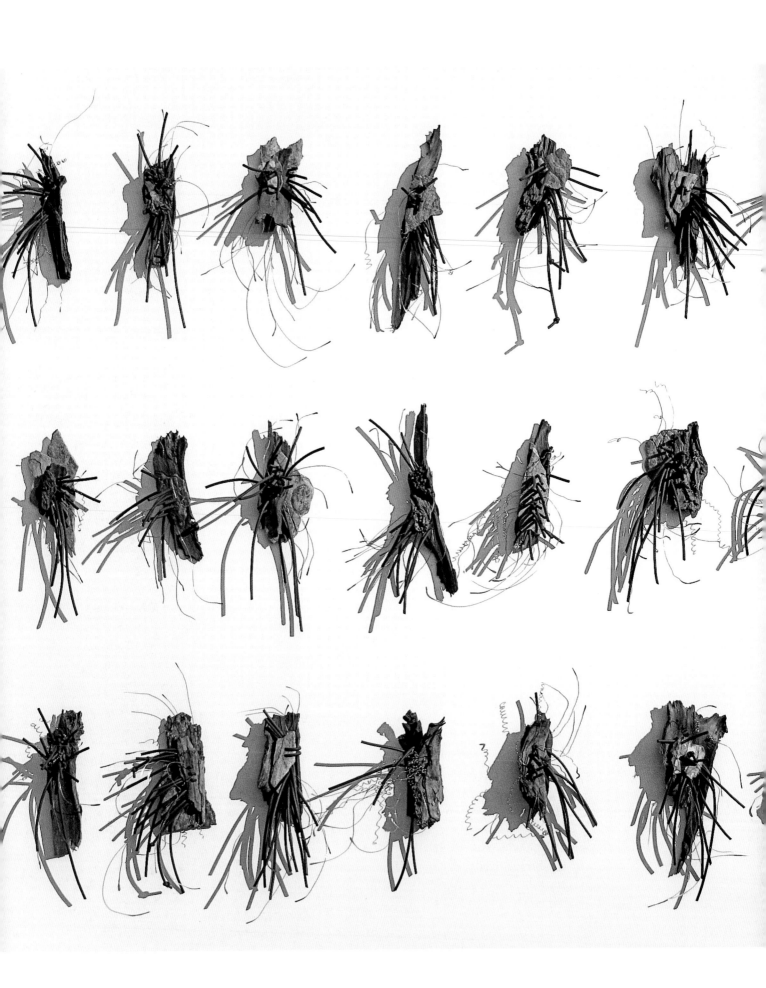

LEFT *Vestiges*; 30 small elements constructed from bark fragments, neoprene cord and copper wire. An example of creating a piece from multiple units (W150 x H100 x D6 cm / W59 x H39¼ x D2¼ in)

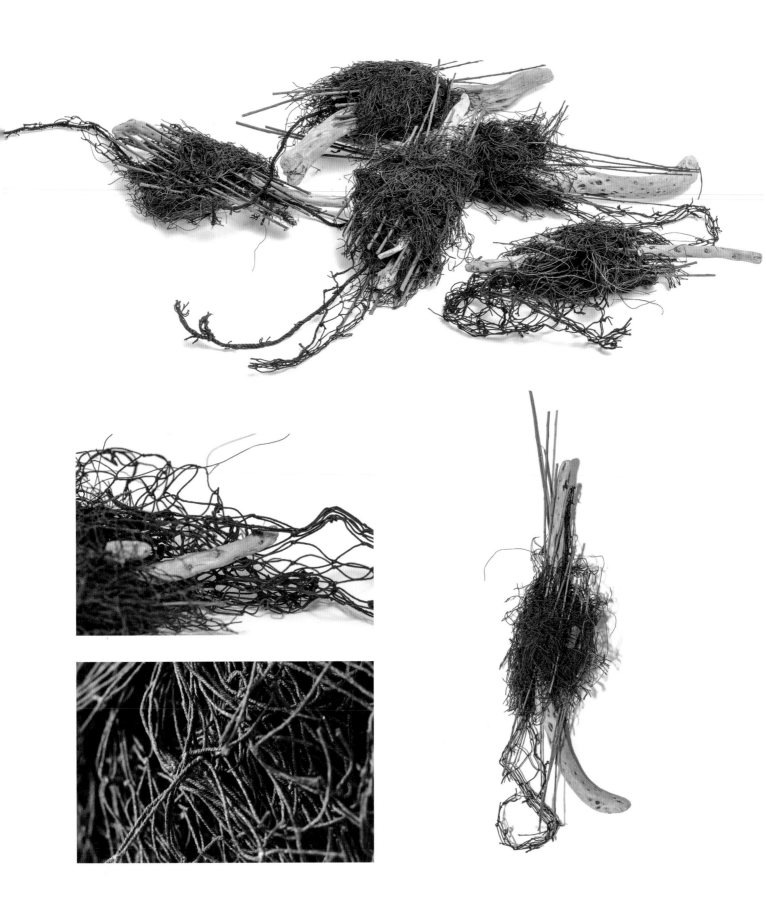

Vestiges

Vestiges are defined as traces or remnants of something that is disappearing or no longer exists. The yew bark fragments are examples of vestiges as they are the traces of a tree which no longer exists.

Idea to use manmade materials to combine the bark fragments to create new habitats
Materials a variety of different 'found' bark fragments, neoprene cord, copper wire
Techniques drilling, assembling, threading and knotting

Vestiges is an installation of thirty small assemblages constructed from fragments of bark. Some are pieces of the ancient yew bark, which are combined with a variety of different bark fragments found amongst undergrowth in local woods.

Symbiosis

At the same time that I collected the yew bark, I sourced a tall, vertical section from the trunk of a laburnum tree. I had no idea what I would do with it but I was drawn to the pattern of its grain.

While researching the theme of human impact on nature I noted the word 'symbiosis' and its meaning: 'relationship between two different species that live close together and depend on each other in various ways.' This evokes the relationship between nature and humans. The notion of symbiosis influenced the way I used the trunk to interpret the theme.

Idea to use textile materials to 'mend' the tree

Materials section of laburnum wood, plastic bottle fibres, neoprene cord
Techniques stitching, insertion

Symbiosis is a comment on deforestation and ongoing conservation efforts. Intervention and an attempt at mending is suggested by the large neoprene cord stitches over gaps in the wood and inserted fibres made from recycled plastic bottles in the cracks and holes.

Plastic pollution and climate change continue to be major global concerns. Researching these themes has expanded my understanding of the problems and has become a vital consideration in my practice.

Self-directed theme: Ebb and Flow

I undertook a lot of research to try and narrow down how I could respond to this theme. I visited the Fishing Heritage Centre in the port of Grimsby, and had a tour of a fishing boat. In doing so, I gained practical knowledge of the fishing industry.

A combination of ideas arose from this to give me direction for the resultant work. I discovered that the fishing industry had been in decline for many years and that cotton fishing nets had been replaced by environmentally unfriendly monofilament nets. The discarding of these old nets became a metaphor for the declining fishing industry. The following piece evolved from this research.

Discarded

This piece consists of five bundles of abandoned items relevant to the decline of the Grimsby fishing industry.

LEFT *Discarded*; 5 units, vintage cotton fishing net, driftwood, lead weights, fragments of rusted metal and wire (W25 x H90 x D20 cm / W9¾ x H35¼ x D8 in)

The form was inspired by research into ghost nets. These are lost or discarded nets that continue to trap fish, collect debris and endanger marine life.

Idea to use discarded materials as a metaphor for the decline of the fishing industry
Material vintage fishing net, driftwood, lead weights, rusty metal, wire, sticks
Techniques cutting, knotting, wrapping

I sourced a vintage cotton fishing net online which I deconstructed and reworked to reflect the changing focus of the fishing industry.

Responding to other artists' work

You can take inspiration for a theme by responding to other artists' works, even if you don't know them personally; for example, responding to famous artists or artworks.

I have recently been collaborating with two other artists, Jae Maries and Shuna Rendel. The following pieces show how I have responded to their artwork.

Biform

Jae Maries is a textile artist who makes large two-dimensional, stitched fabric and painted works. *Biform* (page 86) is my response to one of her pieces.

The artwork being responded to is an abstract view of winter fields in contrasting light and dark colours. Couched threads meander over dark painted sections. I chose to develop work based on the visual aspects of the piece.

Idea a visual response – interpret a section of the two-dimensional piece as a three-dimensional structure or relief
Materials broken fence panels, Indian ink, willow, neoprene cord, nut and bolt
Techniques painting, twining, assembling, trapping

The black and white colours reflect the stark winter landscape and the broken fence panels represent the dark fencing on the artwork. The white willow is held together by neoprene cord representing couched threads. It curls around the form and is clamped in place between the wood sections. The whole construction is held together by a large nut and bolt.

I decided to make another construction mirroring the first, so that combined they would make a larger piece. I chose to hang them on the wall with each a different way up, emulating Jae's process of forming her pieces by cutting them up and reassembling them.

Mutation

This was a response to another piece of Jae Maries' work, this time four small, collaged drawings on paper. The collages were entitled *Regeneration*. They had slashes and raised cuts across the surface, which made them look as though the drawing was trying to break free of the paper.

Idea to interpret the drawings in three-dimensions
Materials willow, recycled paper pieces from a previous artwork, raffia, paper yarn
Techniques tearing, tea dyeing, bonding, assembling, manipulating, improvised technique

RIGHT *Mutation*; willow, recycled papers, raffia and paper yarn (W100 x H60 x D30 cm / W39¼ x H23½ x D12 in)

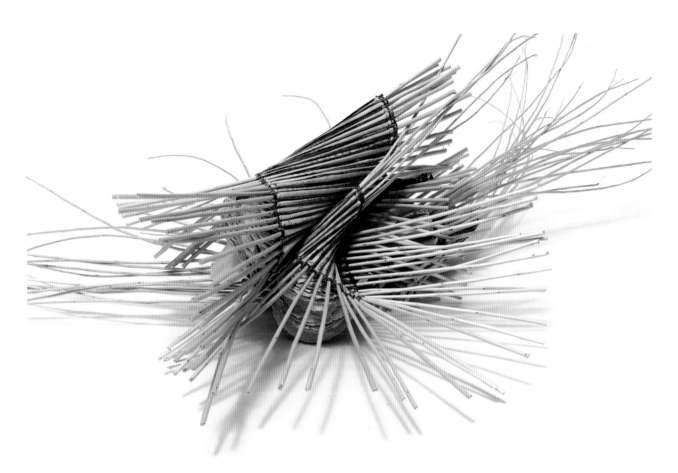

I used willow to represent the slashes. The curved raised cuts reminded me of elements from a paper sculpture I had dismantled due to damage. Inspired by the title *Regeneration*, I reused these papers to make the piece that I titled *Mutation*, as it had mutated from a previous artwork.

Assemblages

Assemblages is my response to a third piece of artwork by Jae Maries. The surface of her artwork is built up from scraps of fabric, stitch and paint. After examining the way that her piece had been constructed I could see that the paint obscured what lay beneath.

Idea to use the words 'obscure' and 'erase' to influence the treatment and juxtaposition of scraps
Materials recycled mixed media scraps including waxed fabric, plastered scrim, painted papers, painted fibres, concrete, burnt wood, painted wood, wire, thread and bristles
Techniques collecting, assembling, wrapping, hand stitch, couching, painting, waxing, burning, plastering

Synonyms of 'obscure' and 'erase' revealed various processes for treating materials to obtain the desired result:

• Cancel	• Fray	• Rubbing out
• Scrape	• File	• Scratching
• Scour	• Erode	out
• Scrub	• Plaster	• Scuff
• Grate	• Wear down	• Shave

The following words also informed the work:

• Obscure	• Hide	• Mask
• Conceal	• Make opaque	• Screen
• Make indistinct	• Make vague	• Veil
• Make faint	• Block	• Wrap

Sampling of these processes freed up my mind to play with the resulting materials, trying different arrangements until I felt something worked. After experimentation, small groups of materials were assembled using textile techniques to hold the disparate materials together.

Entangled

This is my response to a sculpture created by Shuna Rendel. Her piece is constructed from two natural materials: flat even chair cane and curly rounded vine tendrils. The form resembled a 3D linear drawing. The method of construction, in combination with the contrasting qualities of the materials, creates tension which produces different densities of line and open spaces.

Idea to make a three-dimensional line drawing using linear industrial materials to contrast with the fragile, natural materials of Shuna's sculpture
Materials steel reinforcing bars (rebar), iron wire, iron found object
Techniques bending, layering, lashing

RIGHT Small assemblages; wood, concrete, painted bristles, painted paper, waxed paper, scrim and wire. Each layer partially obscures what lies beneath and wrapping holds the disparate materials together

I made two steel and iron drawing sculptures. Initially the layered rebar was going to be fastened together with strong thread but the structure was too unstable, so I used wire to lash them together.

Finding inspiration for themes

- Think about subjects that interest you or that you'd like to know more about.
- Explore aspects of your city, town or village. For example, their history, geography and industries.
- Discover museums: costume, natural history, science, ethnological, zoological, geological, industrial – so many options that could inspire a theme.
- Choose a word or concept to explore.

The following chapter, Conducting Research, will explore research methods. This will help you develop work beyond your initial response.

RIGHT *Entangled*; 1 of 2 units,
reinforcing bars (rebar), found iron object
and wire

Shuna Rendel

Shuna Rendel trained as a sculptor under Anthony Caro. She uses a range of traditional textile techniques to create three-dimensional flexible sculptures. Her studio is overflowing with dried plants, chair cane and experimental samples. In the artist's own words:

'I have always worked with natural fibres collected from the fields and hedgerows around me. A feeling of movement and energy is an essential feature in my work.

The collaboration between Jae, Ann and I has changed my approach to my work considerably. Until we began this, I had focused on flexible structures often using netting and linking, which could take on other forms, such as a hammock (which takes on the form of a body). I pushed these structures to create more rigid forms, developing my own techniques to do so.

Using a piece of another's work as a starting point has led to a very different approach, beginning with a thorough verbal and visual analysis of the work and of the concept behind it, which in turn led to careful selection of materials. That becomes the most important factor in conveying the concept. Finding or developing a suitable technique comes later. The form develops from the concept with the characteristics and possibilities inherent in the material. Rather than sketching ideas, my 3D samples are done to test the scale and structure of material combinations, seeing how their qualities and characteristics work together. The balance of the chosen materials is crucial because the slightest change in weight or scale can completely change the aesthetic.

The Shape Springs Free is my response to Ann's rebar piece, *Entangled* (page 22). I started with a verbal analysis under categories such as lines, shapes, sizes, edges, surfaces, attachments/joints, colour and qualities of materials. I was particularly struck by the granular surface of the concrete embedded in the rusty metal. I then made plant papers from montbretia, bullrush, onion skin, palm fronds, pond iris and apple and willow leaves. These gave me the variety of qualities and textures I was looking for. I wanted the piece to be light and airy, picking up that the spaces between the rebar lines are open and setting off the strength of the rebar. I chose a shape from one of these spaces to develop, manipulating the papers and attaching them to old mattress springs to reflect the rusty spring in Ann's piece, and found the energy, movement and tension I was looking for.'

LEFT *The Shape Springs Free*. Handmade plant papers, old mattress springs (W40 x H28 x D40 cm / W15¾ x H11 x D15¾ in)

2

Conducting research

Conducting research

No piece springs fully formed from the artist's forehead. Many pieces require numerous sketches, maquettes, drafts, or whatever are the interim forms that precede the finished work

MATTHEW LOHDEN

Once I have decided on a theme, I conduct research. I follow a method that I have settled into over many years. My methods are just one approach; other artists will have their own ways of developing work that will be different to mine. The quote by Matthew Lohden, a Hypothetical artist, mentions processes which most artists undertake in order to produce a resolved piece of work.

I hope some aspects of my creative process will be useful when developing your own work.

Background information

There is no replacement for finding what knowledge is already out there. I thoroughly research the subject I want to investigate by gathering as much background information as possible from wherever I can find it. I often take weeks to do this; it becomes all-consuming for a while. I collect articles from the internet, gather images and read books on the subject.

Thorough research helps to find a starting point. The information gathered invariably leads me onto paths of discovery where I find different aspects of the subject to investigate. I once collected images and articles in poly pockets in lever-arch files but now I collect and store the information on my laptop. Information from books is scanned to my computer and combined with information from the internet in a folder with subfolders for each aspect. It is readily available and, with a minimal degree of organisation, allows me to revisit or update it later.

My research falls into two loose groups: collecting words and collecting images.

Collecting words

An example of how I use my collected information is illustrated by my research on the environment. Online research led to many articles on how human activity is affecting nature, causing loss of biodiversity, loss of habitats, and species extinction, with the main cause being deforestation. Researching deforestation and biodiversity opened up a huge area, leading me further into other aspects to investigate, such as changes in land used to grow crops, to create grazing land, road building, airport expansion, for housing and the construction industry, and for quarrying, oil and mining.

While reading the information, I compiled lists of relevant phrases and words, looking up definitions and synonyms as I went. This generated more ideas and concepts to explore. The words and phrases indicated ideas for materials to use or methods of construction. For example: *concreting* over the landscape, *fragmentation* of habitats, *slash and burn*. Printing the lists and pinning them to a board in my studio helps me focus on the project.

Researching the word 'biodiversity' led me to discover the Millennium Seed Bank Project at the Royal Botanic Gardens, Kew, which collects and conserves seeds from all over the world, now housed at their wild botanic gardens at Wakehurst, near

Hayward's Heath in Surrey. I purchased a copy of *Seeds: Time Capsules of Life* by Rob Kesseler and Wolfgang Stuppy. This is full of stunning, coloured images of individual seeds taken with an electron microscope. The images reveal the seeds' delicate, fragile forms constructed from intricate, tessellated segments. Close examination of these images inspired the method of construction of the sculptural forms in my environmental pieces.

Below are some of the words I have collected on the theme of human impact on the environment. Many of these have been used as the inspiration behind, and titles of, my pieces.

- Vestiges
- Relics
- Remains
- Traces
- Under threat
- Consequences
- Ramifications
- Extinction
- Endangered
- Hanging by a thread
- Repercussions
- Considerations
- Implications
- Critical nature
- Trade-offs
- Enveloped
- Engulfed
- Exposure
- At stake
- Up against it
- On the brink
- In bits
- On the edge
- At risk
- In peril
- Encroaching

Below are some of the words I have used as inspiration on the 'boundaries' theme:

- Bound
- Hem in
- Wrap
- Enfold
- Bandage
- Blanket
- Swathe
- Knot
- Lace tight
- Tie
- With strings attached
- Pinned down
- Tether
- Resist
- Cushion
- Repair
- Screen
- Tourniquet
- Drawn tight
- Flatten
- Cramp
- Pinch
- Nip
- Huddle
- Crowd together
- Order
- Boxed in
- Insert
- Wedged in
- Sandwich
- Embed
- Cocoon

Collecting images

My visual research runs alongside the collection of words. I prefer to undertake this through photography, especially close-ups. I also use sketchbooks, but not in the traditional way. Rather than drawing, I use them to jot down ideas, make rough sketches of possibilities, and keep notes of what works and what doesn't.

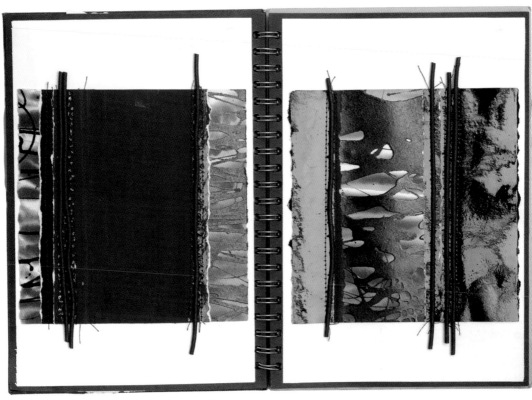

ABOVE AND OPPOSITE Sketchbook pages showing collages and photocopies developed from initial photos of hedges. Used to explore design possibilities

When I was researching the theme of 'boundaries', I took extensive photographs of fences and hedges in the landscape. I then zoomed in, cropped and enlarged images to reveal hidden details. As I am interested in three-dimensional form, I used photographic software to change the coloured image to black and white in order to reveal the structure. After printing these black and white images I photocopied them for a starker effect and to distance them further from the original image.

The boundary photocopies were then interpreted through collage to develop design ideas for construction. Papers coloured with matte and gloss black paint were contrasted with gestural paint marks on white paper. Initially I explored structure, light, weight, density and opacity. I considered whether the weight was at the bottom or top, and whether the light was completely screened, whether it might filter through or where there were clear spaces – this was to help with structural decisions that would determine the use of materials.

As I collect all my information, I spend a lot of time thinking and mulling things over. Ideas seem to pop into my mind at random times, often in my sleep.

ABOVE Small collection of plant materials for inspiration

3

Experimenting with materials

———

Experimenting with materials

What is communicated is the sense of a deep affinity with the material; the need to think, to work out ideas, by making, by engagement with the material; the need to work with the material towards new forms and new visual vocabularies

PAMELA JOHNSON

Most of my work combines textile elements, such as loose fibres, scrim, paper, calico, paper yarn and linen thread, with contrasting materials that are not typically associated with textile practice. These can include concrete, wood, lead and bark. It is through materials that I interpret ideas and introduce significant elements of meaning into my pieces.

It is difficult to describe what draws me to certain materials. It is a visceral feeling; one that cannot be easily put into words. Surface qualities, texture and sensuousness are important characteristics. I am interested in the wabi-sabi aesthetic of finding beauty in the understated and imperfect; I am drawn to materials in their natural state such as cotton fibres and bark or found objects that show evidence of previous use.

There is also a more practical reason why I often choose to work with particular materials for specific pieces. As an artist focusing on 3D construction, it is vital to be able to bring the appropriate types of structural elements into the mix to enable the realisation of the imagined construction.

Like many textile artists, I have cupboards and shelves full of materials and found objects all waiting to be used. I rarely throw anything away because I can guarantee that if I do, it will be the one thing I need for my next project.

After researching the theme, the next step in my creative process is to choose the materials that best suit my idea. I find that hands-on investigation is an important step when trying to assess the potential use of materials that I haven't worked with on a previous project.

Experimentation helps me to discover intrinsic qualities of the material that were not obvious to me initially. In creative embroidery classes, we were encouraged to distress fabrics by burning, waxing, rubbing, tearing, scratching, piercing, fraying, slashing, and so on, to reveal how they change character. I continue to use these processes, and any others I can think of, on the materials I come across or am contemplating using. This includes wood, metal, paper, concrete, slate, found objects, loose fibres, threads and bristles. I will also experiment using hammers, saws and drills, especially on hard media. I am interested to see whether the materials suggest ideas before and after distressing them, and whether other qualities are revealed.

All these investigations help me to gain understanding of the material at a deep level and to ultimately feel at ease utilising it in my work.

I decided to juxtapose materials that show the contrasting qualities of fragility and strength. I tried experimenting to see if I could make fragile materials look strong and strong materials look fragile.

Theme to juxtapose the qualities of fragility/strength
Idea use material investigation to explore these binary opposites
Materials iron powder, variety of fibres, diluted PVA
Technique felted paper making

RIGHT Juxtaposition of natural/industrial materials

Recording results make notes in a sketchbook and store labelled samples for future reference

Making felted paper from silk and plant fibres is one of my favourite textile processes. I use diluted PVA to bond the fibres together to construct delicate forms. Iron, typically associated with strength and durability, was the perfect contrasting material. I discovered that iron can be purchased in powder form, and experimented with the results of combining fibres and iron powder by mixing them in felted paper.

I made many samples, keeping notes on the type of fibre, ratio of iron powder, and strength of PVA dilution for each sample. After a few days I could see that rust was starting to show. Some samples hardly changed; on others the rusting effect was strong.

The best results looked like pieces of delicate, fibrous rusty iron. While the texture and colour are visually compelling, most of the material is hard, brittle and needed gentle handling.

Extending the concept of textiles

When most people think of textiles, they think of fabric and thread. The properties that all fabrics have in common is that they come in long, wide lengths, are flexible enough to bend and can be stitched into. Threads are long, thin, flexible, and come in continuous lengths.

With a little thought, you can push the boundaries and reduce the textile property list to a flexible sheet. This could include copper sheeting, lead sheeting, paper, cardboard or rubber sheeting. Similarly, threads could be categorised as anything long, thin and bendable; this could include wire, paper yarn, cord, monofilament or chair cane. This opens so many more materials for use.

One benefit I have found from this expansion of the concept of textiles is that some of these new materials can provide much better structural elements for my 3D construction than was possible with traditional possibilities.

Here are some of my favourite thread alternatives:

Rubber

The rubber I use is neoprene cord. I first discovered this among a box of rubber samples I had requested from a manufacturer. Neoprene is a general-purpose, sponge cord used in various sealing applications and is sold by the metre.

I am drawn to it for its tactile qualities and its potential to evoke different connotations depending on its use. It has the following properties:

- is interestingly flexible
- is smooth and sensuous
- it comes in different thicknesses
- is synthetic so doesn't disintegrate like natural rubber
- has different characteristics depending on if it is cut short or left to hang
- regains shape if squashed for a period then released

- is slightly stretchy
- is versatile; it can be used for stitching, weaving, knotting, or holding constructions together

Bristles

I first thought of using bristles when I was trying to create spiky seed forms with handmade silk paper. For spikes, I experimented with stiffened thread, small thin pieces of willow, and fabric-covered wire, but none looked right. I then cut some bristles off a household brush and incorporated these into the paper. They were strong and stiff enough to carry the weight of small fibre forms while still looking delicate.

I find these useful for their following properties:

- short
- stiff
- delicate
- spiky
- fine or coarse
- natural or manmade
- different colours and lengths

Wire

Wire is a very versatile material and is especially useful when making three-dimensional work. It has the following qualities:

- is analogous to thread
- can be used to stitch, knit, crochet or weave
- can be used sculpturally
- is easily manipulated
- holds its shape, so can be a structural element
- can be hammered and flattened
- able to be cut with scissors or wire cutters if thin and with bolt cutters if thick
- comes on reels or as pre-measured lengths
- can be dyed if it is covered in cloth or paper

Wire is readily available in different thicknesses and metals from DIY shops, garden centres, floristry suppliers, online sellers, craft shops and jewellery suppliers.

Ordering wire can be confusing as the gauge system of sizing wire is the opposite of what you might expect. The higher the standard wire gauge (SWG) number the thinner the wire:

Low number SWG = thick diameter wire
High number SWG = thin diameter wire

Below are the wires I have used:

- **Binding wire** available from floristry suppliers. It is made of mild steel, comes on reels and is easily manipulated.
- **Stub wire** available from floristry suppliers. It comes in pre-measured straight lengths, not on a reel. It is useful if you want a wire to stay straight.
- **Galvanised wire** this is a silvery coloured steel wire with a zinc coating to prevent it from rusting.

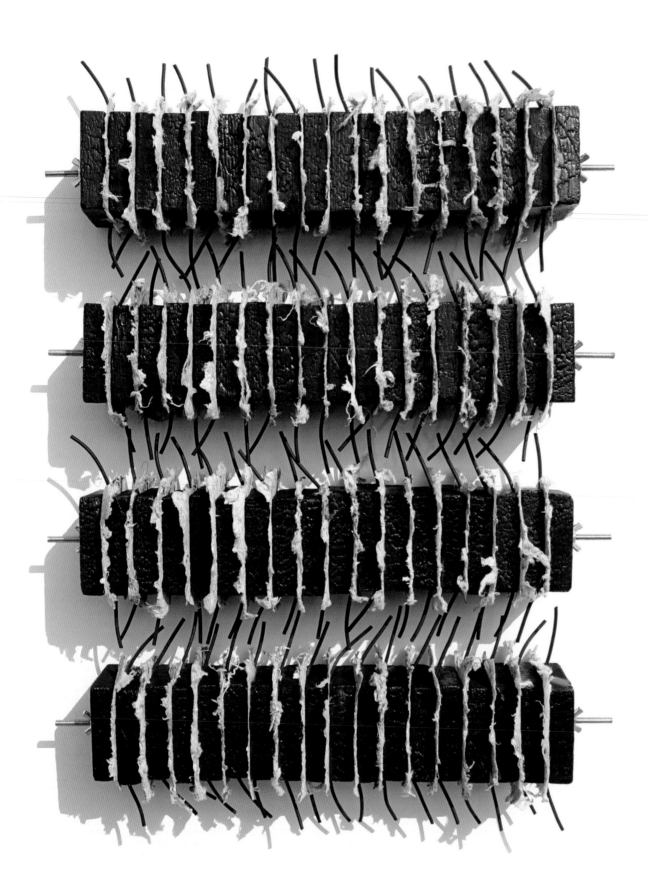

ABOVE *Blocked Off*; burnt wood, felted silk paper, neoprene cord, threaded rods, wing nuts and washers. An example of creating a piece from multiple units (W50 x H70 x D8 cm / W19¾ x H27½ x D3½ in)

It is weatherproof. It can be more difficult to manipulate than other wire.

- **Copper wire** very easily manipulated, depending on gauge. Available on reels, in a loose coil or as long rods. It changes colour when heat is applied.
- **Paper-covered wire** thin gauge, can be dyed or painted. Available as straight pre-cut lengths.
- **Cotton-covered wire** comes on reels in a continuous length. Can be dyed or painted. The cotton end may unravel when cut – a little bit of PVA rubbed over the end stops this.

Unconventional materials

While pushing the boundaries of tradition, I like to mix my extended range of textiles with materials that are most definitely not textiles. In fact, I generally prefer to make the contrast as stark as possible. These materials give me the structural elements that are so necessary for 3D construction. Here are some of my most used contrasting materials:

Wood

David Nash is an internationally acclaimed wood sculptor and land artist. He uses chainsaws and blowtorches to create large-scale sculptures. Seeing his burnt and blackened forms had a profound influence on my work.

Wood was one of the first non-textile elements I combined with textiles. I started by making my own wooden beads from lengths of new wood and progressed from there. I now prefer to use wood that has a history: reclaimed planks, old broken fence panels, driftwood, and any pieces with nails, paint marks, holes, cracks. The more imperfect the better. But as I didn't always have pieces like this handy, I experimented with ways to change the appearance of the new wood.

My solution was to burn the wood with a blowtorch to darken it. After removing the powdery surface with wire wool and wiping it with a damp cloth, the wood underneath was a smooth, dark brown.

I discovered that by holding the blowtorch for longer against the wood the surface starts to contract and split from the heat, leaving a black, velvety, crackled surface and softened edges.

The blackened surface flakes off quite easily, but it is possible to retain the crackled surface without damaging it by carefully applying clear matte wood varnish to seal it. The effects of this can be seen on *Limitations 1* (page 66) and *Blocked Off* (page 42).

I have since discovered this is an ancient Japanese technique called shou-sugi-ban. The traditional method uses several layers of tung oil rubbed over

Safety

Burn the wood outside wearing a dust protector face mask, safety glasses and thick gloves, to shield hands from burning and prevent breathing in the carbon dust.

the surface to seal it rather than varnish. The oil penetrates the wood beneath to weatherproof it.

It is possible to buy ready-charred wood cladding to cover houses. There is even faux shou-sugi-ban cladding available that hasn't been burnt at all. Where's the fun in that!

Concrete

When I first considered using concrete, I did a great number of experiments trying to make my own mix of sand, cement and gravel. Getting the correct consistency for the mix was trickier than I expected. In the end, I resorted to ready-mixed concrete, which comes as a dry mix in bags and just requires the correct amount of water adding. The important aspect for me was that the material should convey meaning rather than illustrate how good I was at making my own concrete mix.

To mix small quantities all I needed was a plastic bowl, an old spoon and a jug for the water. It's important to put the concrete mix in the bowl first and gradually add water to it, stirring as you go to incorporate all the powder. It will usually take less water than expected to get the correct consistency.

I tried sampling by making my own small rectangular moulds from pieces of wood attached to a flat base with masking tape. After filling the moulds with wet concrete, I made imprints with objects and fabrics that were coarse enough to reproduce well in the concrete's surface. If the concrete consistency was too runny, small objects and pieces of fabric were

Safety

Concrete contains chemicals and compounds that can cause skin irritation. Concrete dust can be harmful to eyes and the respiratory system, and wet concrete can cause chemical skin burns. To keep safe:

- Wear waterproof gloves, safety glasses and a face mask.
- Cover bare skin by wearing suitable clothing.
- If any concrete touches your skin wash it off immediately.
- Take care of your back if moving heavy bags of concrete mix.

completely swallowed up when the mix was poured over them.

I discovered polystyrene will peel off hardened concrete, so pressed nails and similar items into a sheet to replicate stitch marks, then removed the nails and poured concrete over it. When cured, the polystyrene was peeled off, revealing raised stitch marks in the concrete.

I also tried using sections of plastic gutter as moulds, using both the convex and concave sides. The concrete that was against the plastic of the gutter turned out beautifully smooth and sensuous, while the surface of the concrete open to the air was rough and would have been unacceptable in an actual piece.

LEFT Experimental concrete sample

When clearing up after using the concrete it is important not to pour anything down the sink. I leave leftover concrete to dry on a suitable surface and dispose of it at a recycling centre. I wipe clean any containers used for mixing concrete with paper then fill them with water and scrub them clean outside the studio to make sure they are reusable.

Lead

I first came across this material in connection with textiles when I saw lead beads in Indian embroidery. I bought some lead from a builder's yard and received some very strange looks when I told them I wanted it for embroidery.

Safety

Lead in its metallic state is not absorbed through the skin. Your body only absorbs lead when you:
- breathe in lead dust, fume or vapour
- swallow lead by eating, drinking, smoking, or biting your nails without washing your hands and face

If hammering lead:
- Wear gloves for extra protection and wear a face mask to prevent breathing in lead dust.
- Wash any exposed skin.
- Avoid working with lead if you are pregnant. An unborn child is at particular risk from exposure to lead, especially in the early weeks.

The lead in the roll was about 1.5mm (1/16) thick, so needed to be hammered to make it more malleable. I wore safety glasses, gloves and a face mask while doing this. The lead changed its character the more it was hammered; I discovered that it picked up the hammer prints and also the texture of the surface being hammered on, which in my case was a concrete floor. It spread out and became thinner as it was hammered, and I soon discovered it was possible to make very delicate shapes. The surface being hammered is quite silvery to begin with but soon dulls to a matte grey.

Rebar

The word 'rebar' is short for reinforcing bar. It is made of steel and is used in the construction industry to reinforce concrete and masonry structures. You often see rusty piles of it lying around building sites or protruding from foundations before buildings are erected. It comes in individual bars or as a large-scale mesh. The surface of the bars has a ridged pattern along the length to enable the concrete to grip.

It may seem like a strange material to incorporate into my work, but it is an invaluable reference to the construction industry in many of my pieces including *Secret Support* (page 100), *At Stake* (page 99), and as a linear material in *Entangled* (page 22).

I obtained the reinforcing bars by visiting a rebar manufacturer to salvage any offcuts. Telling them I wanted to make a corset out of it for an art exhibition caused great amusement, but they kindly let me take any pieces lying around their grounds.

RIGHT *Boundary Lines*; hard materials – wood, slate, metal mending plates, variety of screws, washers, nuts, bolts; soft materials – neoprene cord, paper yarn, scrim, felted silk paper (W72 x H107 x D10 cm / W28¼ x H43¼ x D4 in)

Wire metal mesh

Metal mesh is a very versatile material. It resembles tapestry canvas, netting and scrim. It can be used in the same way as other mesh fabrics, but being metal it is useful as a structural element in three-dimensional work.

I have used basic wire mesh such as chicken wire and mesh panels from animal cages as fencing material for my boundary-inspired pieces. Chicken wire is made from thin, flexible, galvanised steel and comes in a roll. Mesh panels for animal cages are more rigid as they are manufactured by welding thicker wires to form rectangular or square holes. They can be cut to size with wire cutters or multi-purpose heavy-duty scissors. The cut ends are very sharp so care needs to be taken. Both types are readily available from DIY shops and garden centres.

There is a huge variety of alternative metal meshes available online. These have varying sized and shaped holes and come in different metals, which could be inspirational for alternative themes.

Slate

I use a mixture of old and new grey slate roof tiles. The old slate came from a builder's reclamation yard. The tiles are available in different sizes and are remarkably strong considering their thickness.

I chose tiles because I wanted flat pieces to use on my boundary-themed pieces to reference the slate fences in North Wales. Some slate fences are constructed from huge irregular slabs of old slate, fastened loosely together with wire. Others are newer with smaller regular shaped pieces connected by wire.

Slate can be cut using an electric tile cutter. Drilling holes can be difficult as the weight of the handheld electric drill invariably makes the slate break. I tried a masonry bit without success but eventually solved the problem by lying the slate on a piece of flat wood then drilling through the slate and into the wood. This prevented the sudden drop when the drill completed the hole, so the slate remained whole.

Natural materials

Andy Goldsworthy is an environmental artist. His love of nature is reflected in the extensive range of natural materials he works with, from small, delicate leaves and thorns to monumental work using dry stone walling. He records his ephemeral work through photography. His constructions are built by combining many small elements to create a larger whole and he often revisits the same form in different materials and scale. Discovering his work had a huge influence on my own approach.

Living in the countryside I am lucky to be surrounded by interesting natural plant material that can be incorporated into artwork. I use these materials for their natural qualities: their texture, twisted and delicate forms, and muted colour. They are a constant source of inspiration. I prefer to work with them in their natural state.

RIGHT *Hanging by a Thread*; one unit from a series of five; brown paper packaging, willow, neoprene cord, linen thread (W30 x H65 x D17 cm / W12 x H25¾ x D6¾ in)

Willow

I use willow as a structural element in my work. Initially I dismantled a willow garden screen to access the rods, which also provided me with a quantity of wavy wires that had held the rods together. Garden screening is a boundary marker so I could use it in my boundary-themed pieces. Later I purchased bundles of willow in different shades and lengths. Also, a friend gave me some reddish-orange willow that she grew in her garden.

I have found the 90–120cm (3–4ft) lengths more useful than the longer rods as they are thinner and have delicate ends, which is what I prefer visually. Being thin they are easy to manipulate, can be cut with secateurs, and can be fastened together to make longer lengths if necessary. Sometimes I leave the rods straight, as can be seen in *Adrift* (page 92).

By contrast, in the *Hanging by a Thread* series (page 49), the rods were first soaked in the bath until they could be gently bent into the desired shape. If using rods longer than your bath, plastic soaking bags are available from willow growers.

Soaking times can vary from hours to days depending on the type of willow. I tie the rods to hold them in position and hang them to dry. The tie can be removed when dry and the willow retains the shape and is ready to use.

Bark

The ancient yew bark previously mentioned on page 15 inspired me to collect bark from other trees while walking in the local woods. I never remove bark from growing trees, there are always plenty of pieces lying in the undergrowth that have dropped off dead trees. The colours and textures vary hugely. These fragments have been used in combination with the yew bark to make *Vestiges* (page 19), the *Habitat* series (page 91) and *At Stake* (page 99). I gently brush the pieces to remove any insects and dirt before storing them in boxes until they are used.

Fibres

Over the years, I have amassed a huge quantity of loose fibres, each with different characters and qualities: silk noils, cotton noils, silk throwsters waste, silk cocoon strippings, flax, hemp, and recycled plastic bottle fibres. I prefer to use the fibres loose, or to create delicate forms from felted silk and cotton papers to contrast hard materials.

Paper

Another favourite material is paper, especially sustainable, handmade Himalayan papers.

- **Himalayan papers** Nepalese mitsumata and lokta papers and Bhutanese resho paper, are all made from the inner bark or bast fibre of plants that grow above 3000 metres (9842 feet) in the Himalayan foothills. I prefer to use these as they are eco-friendly; the plants are harvested above ground level and can be re-harvested after three or four years so they are a renewable resource. I like

the natural coloured sheets that are firm enough to hold a shape when used sculpturally. Being handmade they have attractive deckle edges. I often wax or distress their surface for texture.
- **Tracing paper** it is stiff enough to be used sculpturally and its translucence is a quality to experiment with.
- **Paper yarn** this is a very versatile material. It can be used as any other thread or yarn but being strong and stiff it is ideal for sculptural work.

Incorporating found materials

Found materials can include just about anything: interesting items in your home, treasures in the attic or shed, curiosities found in skips, on beaches, in woods or on the road, and so on.

Your finds provide you with unusual materials to experiment with and their provenance could inspire a theme to explore. They could also introduce connotations adding meaning to your work. Figuring out how to assemble found objects or incorporate them into work will encourage unconventional solutions to be found.

Several of the featured artists in the book incorporate found items in their work. Michael Brennand-Wood includes a wide range of unusual found objects in some of his pieces; Shannon Weber constructs her sculptures almost entirely from found materials; Elizabeth Couzins Scott repurposes items; and Nerissa Cargill Thompson casts her cement sculptures in litter.

Juxtaposing materials

Ceramist Gillian Lowndes' practice straddled the boundaries of fine art and craft. Her biographer Amanda Fielding describes Lowndes as being 'renowned for her sensitive investigations of material and process, of serendipity and sculptural form'. Her small-scale, quirky sculptural assemblages juxtapose strips of slip-covered fibreglass matting with found household items such as bulldog clips, whisks, can openers, forks, slotted spoons and wire. The fired results are unlike anything I had seen before. Her unconventional approach to materials has had a strong influence on my own attitude to combining and juxtaposing materials.

Another approach to making is to play directly with materials. I have found it is a particularly useful practice if you are struggling with creative block. It can also be a highly productive strategy. Albert Einstein declared that 'play is the highest form of research'.

Influential textile designer and weaver Anni Albers noticed when teaching that 'uninhibited play with materials resulted in amazing objects, striking in their newness of conception'.

Artist and author Austin Kleon suggests that 'new ideas are formed by interesting juxtapositions, and interesting juxtapositions happen when things are out of place'.

You could juxtapose materials with binary qualities:

- hard/soft
- strong/fragile
- flexible/rigid
- natural/manmade
- transparent/opaque

Or, like Gillian Lowndes, combine any items no matter how strange and unconventional.

Playing with materials frees up my imagination. When I played with mixing rebar with natural materials and juxtaposing plastered scrim and inked cardboard, I wasn't thinking about meaning, concepts or theme – I just enjoyed seeing the visual results.

I struggled to find a way in to the making of *Assemblages*, but managed to move forward by playing with scraps from my studio, trying them out together in different arrangements. Eventually the pieces started to come together. Some of the combinations worked, so I carried on making a group of them.

It is difficult to explain how I know if combinations work, or when a piece is resolved. It's an internal feeling; sometimes my heart beats faster. Textile artist Rozanne Hawksley describes it as: 'The instinctive, inexplicable feeling that YES – this is right – this is right.' I felt this when I first started combining textiles with other media. I felt this again when juxtaposing my scraps and knew I had found my way in.

Normally, I try to put constraints on myself to not use more than three different elements in a

piece. However, *Assemblages* (page 22) is made from so many disparate components that I linked them through colour and placement instead.

When working directly with materials, I photograph various groupings and bulky three-dimensional samples as I go to keep a record of progress and to remember alternative ideas. I keep these in my sketchbooks.

Unsuccessful experiments are also useful to keep as they could spark an idea if looked at later. In fact, looking back over old samples, sketchbooks and notes is a great way to remind you of what you have done previously, and it can stimulate different interpretations or ideas for new work.

LEFT Juxtaposing plaster with scrim (hard/soft) and plastered scrim with black inked corrugated cardboard (light/dark)

ABOVE AND OPPOSITE Small assemblages constructed by juxtaposing leftover scraps from my studio

Michael Brennand-Wood

Michael Brennand-Wood is an internationally renowned textile artist. He has forged a unique practice challenging the boundaries of textile and craft approaches. His intricate and eclectic constructions focus on ideas and how to interpret them through materials.

The juxtaposition of stitch and wood in his early grid reliefs made a significant impression on me and expanded my understanding of what embroidery could be. His later works have multi-layered narratives; decorative from afar but revealing a political message on closer inspection. *Babel* is one such piece. Embroidered blooms contrast with a tower of toy soldiers that project 70cm (27½in) from the wall. Michael comments on his work:

'*Babel* is based compositionally on an American arena-seating plan; arenas are ceremonial places where events unfold. The circularity of the work is equally referential to a flower and echoes the head of a flower in form. *Babel* is clearly about the Tower of Babel. The stamen, the phallic form at the centre, is constructed from hundreds of fused toy soldiers, which create a tower as they march upwards. The machined text collage is purposefully nonsensical, yet within the shattered phrases are the odd clear word or phrase, 'lie' being an obvious example. I wanted to create a predominantly black and white work with the odd flake of colour. The blooms at the end of the wires are a form of cap or military badge, graphic encapsulations of campaign history that are sewed or pinned to the body.

I'm interested in pictographic, ideographic floral shapes, which initially appear pretty or decorative when viewed from afar. Closer scrutiny reveals less pleasant associations, such as the use of land mines, skulls or explosive stars. 'Death's head' moths are a symbol of change. Optically, the work visually flickers because of the limited, almost monochrome, palette, apart from the tiny areas of red. Simulacra or shadowy likenesses to familiar objects have intrigued artists and mystics from the earliest times. Everyone at some point has seen faces or figures in the surface of rocks, tree bark, clouds, damp stains or peeling walls. The shadows surrounding *Babel* enhance the transient state of the core imagery as it disappears into the ether.'

LEFT *Babel*. Embroidered blooms, wire, acrylic, collage, toy soldiers, wood base (D90 x D70 cm / D35½ x D27½ in)

Shannon Weber

S hannon Weber is an American textile artist living in Oregon. Her eclectic sculptures and assemblages combine wildly disparate materials, held together with stitch, weaving and binding, combined with her own improvised non-textile techniques. Shannon comments on her practice:

'As an adult I have lived in very remote locations of Oregon and currently reside in a small rural town. I taught myself to construct and weave organic forms using whatever I could find while I was living in a fishing village along the Rogue River. As a curious eccentric type, I am attracted to the odd, the forgotten and things of decay. I see every material as an option. I find myself absorbed by where these materials and items that I collect are located, and the mythologies they each hold from growing wild or laying on the ground where they're found, to drifting in from one place or another on the water. That inspiration drives my work, with the dialogue of each and every piece.

My studio is a lab of sorts. I haul my collections around my studio and stack them around my work bench waiting for their dialogue to begin. Not all material is ready to be used; some have to be coaxed by pounding materials with rocks, setting them on fire, or boiling to provide flexibility or interesting markings from these processes.

I work with 80% locally sourced reclaimed materials and harvested organics of all kinds. A short list would be Pacific NW kelp, bike or tyre rubber, wood, wire, ocean debris and plastics, paper, wire, wax and rocks. I am fearless about trying out new materials. The worst-case scenario is that something won't work for whatever I have in mind, so I have learned something and now I will move on.

My work is designed by forms of tension, by mixing layers of different applied techniques with a variety of materials, layer by layer without the use of glue. Some noted techniques include jewellery-inspired cold connections, stitching, various forms of weaving, wax, and others. With that said, nothing starts the same. I never know for sure what I will have to work with, so I am always reinventing the wheel. The only certainty is the use of some kind of thread for stitching or binding. I seem to be known for a style and a lot of my work reflects boats, or an artefact ambiance of certain objects. This is not directed by me. I am a vehicle in these processes but not the driver. As wild as it sounds, the materials know the way.'

LEFT *Dolly Dock Boat Series* (detail). wood block, reclaimed metal, stick, paint, Hag stone, rope, beach plastics, reed, waxed linen thread, paper, encaustic, Ostrich egg shell, flat clay beads (W61 x H25½ x D28 cm / W24 x H10 x D11 in)

Elizabeth Couzins Scott

Elizabeth Couzins Scott originally studied 3D Design (ceramics) but felt restricted working within a purely ceramic tradition and moved towards textile, collage and mixed media. She is interested in recycling and the transformation of found materials coupled with handmade paper, textiles and stitch to create artefacts and assemblages of disparate components as she explains when describing two of her pieces:

I'm Worth It

'This work is inspired by the dark side of experimental fashion design, whose more conceptual designers have explored the symbolic and cultural meaning of consumer culture and contemporary anxieties. The interpretation of these themes by modern historians such as Caroline Evans by way of mythology, identity and gender allows me to develop ideas into imagery using accessories and articles of clothing.'

Mermaid's Purse (Lost at Sea)

'Environmental decline and degradation are facing our oceans. Climate change and plastic pollution are already threatening our marine life. Artists need to contribute to the growing body of environmentally conscious artworks that carry a responsible message. The legends and mythology of the seas from many cultures tell of people who live in the underwater world, selkies, mermaids and sirens. Thinking of these stories, I began to wonder what if articles lost at sea were collected by the sea people and repurposed into items for their use.

I began to consider a fictitious museum style display that reveals an imaginary mermaid's collection of accessories, everyday objects from our world given a different life. The display would also show her concern for the disappearance of sea life and contain her reliquiae of precious fossilised sea forms that have also been lost at sea or are in the process of disappearing from our oceans. The artefacts I have made to convey this message are all constructed from found debris washed up on the shore, pressed porcelain pieces and cast, recycled paper and textiles, found materials from the seashore, broken seashells, sand and salt.'

LEFT (top) *Mermaids Purse*. Canvas purse, found materials from the seashore, broken seashell, sand and salt (W11 x H12 x D3 cm / W4¼ x H4¾ x D1¼ in)
LEFT (bottom) *I'm Worth It*. Found handbag, distressed polyester fabric, black tacks and barbed wire (W19 x H30 x D10 cm / W7½ x H11¾ x D4 in)

4

Conveying meaning

Conveying meaning

What I am trying to get across is that material is a means of communication

ANNI ALBERS

If you are interested in adding meaning to your work, this section explains how I try to do it. Some questions to ask yourself are:

- Do I want to convey meaning through my art at all?
- If so, how do I communicate meaning to the viewer?
- How literal do I want to make my messaging of meaning?

Not all artists want to convey meaning, of course, but I almost always do. I want to give gentle guidance to the viewer, to get them thinking in the right direction but not be so literal as to prevent them thinking widely around the theme of the meaning. It's a difficult balance to convey the meaning, while being vague enough that the viewer is not only engaged but also free to adapt and expand on it and make personal associations.

Of course, there is only so far that you can go with guidance towards meaning. Some viewers are not interested in such things and are drawn to a piece by the visual aesthetic. The people I aim to talk to are those who not only seek meaning but also benefit from their interpreted understanding of my guidance.

In this chapter I will describe how I:

- incorporate meaning in a piece
- convey that meaning to the viewer

Below I categorise the areas of messaging. However, striving to both incorporate and convey meaning is something that I continue throughout the entire process of making a piece, whether this be choosing a theme, researching ideas for a specific piece, or investigating which materials might be most appropriate. This is an ever-present backdrop to my artistic process.

Theme

This is the source of the meaning for the piece. Researching the theme generates ideas and it is a major source of broad-brush meaning. However, in most situations, consequential meaning is not conveyed directly to the viewer.

The theme is the starting point that provides focus and direction, driving research and revealing potential aspects for further investigation. The information collected takes you further into the theme, perhaps revealing areas to explore and raising issues of which you were previously unaware. Thorough research provides a deeper understanding enabling a meaningful way of making rather than a superficial response to the theme.

Titles and ideas

The idea behind the piece provides a direct and focused source of meaning. In an exhibition, the title of the artwork is often the first direct communication with the viewer. I generally like to make this the first hint at meaning.

Before starting to make something, I consider which aspect of the theme will be behind the piece I am aiming to produce. I usually choose the title first from words or phrases that are already established in my initial research for a theme. The idea and the title are then used to steer the direction of material choice and the making process. As a result, the title, the idea and the meaning I wish to convey are tightly bound together from the outset.

Exhibiting: statements

If you will be exhibiting your work, an artist's statement is usually required to give the viewer the idea behind the work. If you are visiting an exhibition, this is another method the artist uses to pass meaning to the viewer. There is often a limit to the words for a statement: fifty words is common. I generally think long and hard about these, as I find it difficult to condense my thoughts into such a short statement.

Itemising the materials and techniques used provides hints and is essential for my way of working as I try to impart my message through materials and processes. I attempt to make sure that all my communications, verbal or written, re-emphasise and clarify the more subtle messages from the piece itself.

Exhibiting: presentation

How work is displayed is another way to convey meaning. For example, in my installation *On the Brink* (page 69), the work displayed balanced on very thin pins in the wall to convey vulnerability, which is a central message behind the piece.

The case studies on the following pages show how I tried to convey meaning from concept, through design, experimentation and construction.

Conveying meaning case study: LIMITATIONS 1

Theme — Boundaries.

Motivation — To explore the concept and function of boundaries.

Relevant research for the theme — Finding definitions, words and phrases regarding boundaries: a boundary is a line that marks a limit/a dividing line; its function is to control, contain, restrict.

Title — *Limitations 1*

Idea behind the piece and its title — I wanted to capture the closed-in feeling and control that hedges and fences impose on access to the countryside in Cheshire where I live, which was so different to the wide-open boundary-free moors and hills I had been used to when growing up in North Yorkshire.

Exploration and research for piece — I took many photos of the narrow country lanes bounded by hedges and explored their structure through photocopies and collage. I chose signifying words from the 'constrain, control and restrict' part of the theme research: restraining, stiff, unbending, tight, narrow, confining, limiting. These were used in the choice of materials and informed the construction.

Materials — Reclaimed planks, silk noils, threaded rods, wing nuts.

Materials connotations — I chose to use wood as the principal material for its stiff and unbending qualities, enabling it to control and restrict. The silk noil fibres referenced textile practice.

Construction — The planks were sawn to size then burnt and the surface sealed. Holes were drilled to allow threaded rods to be used for the assembly. As each plank was pushed onto the rods, the loose silk noil fibres were spread across the plank, then another plank placed on top, and so on until the top of the rods was reached. Wing nuts and washers were used to hold the construction together at the top and bottom of the threaded rods.

Construction connotations — The process of burning the wood hints at destruction and the breaking down of boundaries. Layering the construction shows distinct boundary lines, with the fibres being trapped and confined between the planks. Wing nuts can be easily unscrewed and were used to imply that the structure, like enclosures, is not permanent but is a manmade construct.

Conveying meaning case study:
ON THE BRINK

Theme —	Human impact on the environment.
Motivation —	Concern for endangered species.
Relevant research for the theme —	Reading articles online about environmental issues such as loss of biodiversity.
Title —	*On the Brink*
Idea behind the piece and its title —	Insects are vital to the health of the planet. However, pollinators such as butterflies and bees are endangered species. This piece highlights the plight of insects and reflects the probability of survival to maturity of insect species on the brink of extinction.
Exploration and research for piece —	One article stated that 40% of insect species worldwide were in danger of becoming extinct in the next few decades due to habitat destruction and deforestation.
Materials —	Ugandan bark cloth, sticks, entomology pins, wire.
Materials connotations —	I could have chosen a delicate fabric for the pupae to connote vulnerability, but I wanted a material that would also reference wood and deforestation. Ugandan bark cloth was suitable as it is carved from trees, its production is endangered due to deforestation and change of land use, and it is a strong fabric that would hold the form.
Construction —	Strips of bark cloth were cut, and the edges burnt and sealed with diluted PVA to prevent crumbling. Burning the fabric was informed by the forest-clearing method of slash and burn. When dry, the strips were twisted and bonded into pupa-shaped forms. The black pupae are the result of a lucky accident. I dyed the strips before forming them into pupae and they shrivelled up and shrank in the dye. After much manipulation, it was possible to create misshapen and irregularly sized forms, which seemed more appropriate to the narrative than my original idea.
Small holes were drilled through the top of each bark cloth element and a fine wire threaded through so the pupae could be attached to the sticks on pins.
The overall form is a wall-based installation consisting of 100 small sculptural pupae forms attached to sticks with entomology pins.
Construction connotations —	The piece reflects the 40% probability of survival to maturity of insect species. The slash-and-burn metaphor infers vulnerability and danger. The black, misshapen and irregularly sized forms are pointers to decomposition of life.
Entomology pins are used by insect collectors. By fastening the pupae to the sticks with entomology pins then balancing the sticks on these pins, I aim to further convey vulnerability and suggest the uncertainty of the insects' survival.

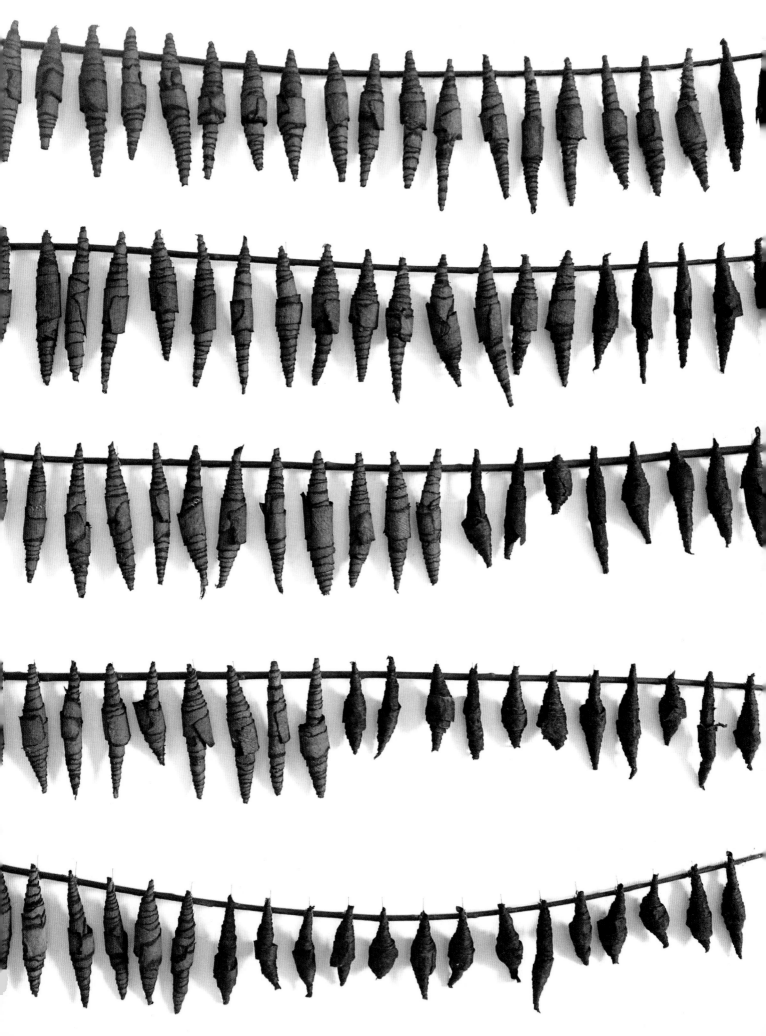

Conveying meaning case study: ECOTYPE

Theme —	Human impact on the environment.
Motivation —	Concern for nature.
Relevant research for the theme —	Finding information on the mass extinction of species.
Title —	*Ecotype*
Idea behind the piece and its title —	Investigate the effects of change of land use and the possibility of species adapting to new conditions.
Exploration and research for piece —	I read that an ecotype is a group within a species having unique physical characteristics genetically adapted to particular environmental conditions. Would this make it more likely to survive or more likely to become extinct?
Materials —	Mitsumata paper, lokta paper, A4 printer paper, paper yarn, tea dye, wax, gesso.
Materials connotations —	The paper was selected from two sources: sustainable papers from plants adapted to living in the Himalayas, and generic A4 printer paper made from wood pulp that involves the felling of trees. Combining them in one form suggests the mixing of genes to create a hybrid species. Would this improve the possibility of surviving?
Construction —	The construction was informed by electron microscope images of seeds. Paper elements were prepared by printing words and phrases from the research on some, texturing the surface of others, then tearing them into shape. The elements were manipulated by changing the shape slightly, then they were threaded onto paper yarn. The variety of size and order of threading the elements determined the finished form.
Construction connotations —	The large-scale structure references a seed, magnified. Perhaps it has grown as a result of its hybridity. Unfortunately, this piece was later damaged and has since been dismantled. Parts of it have been recycled and reused in another piece.

5

Preparing the elements

Preparing the elements

I like things that are labour intensive. You make a little thing and another little thing
and eventually you see a possibility

KIKI SMITH

In this chapter I will look at the preparation of the various elements required before I construct my pieces. The word 'elements' in this context refers to the individual sub-components that I combine to create a finished construction. To prepare these elements I often use tools that are not traditionally associated with textile practice.

Tools and techniques

When I started embroidery, the only tools I expected to use were a sewing machine, a selection of needles, and scissors. I never dreamt that my practice would lead to me needing so many other tools. Initially I used what we already had in the house, but over the years I realised it saves a lot of time and hassle if you have the correct tool for each job. I have since systematically upgraded my tools as need arose.

Below are some suggestions for tools which you may find useful if using hard materials.

Hammers
There are many types of hammers but the following three are ones that I have found most practical:

- **Ball pein hammer** lightweight. It has one round side and one flat. The flat end is for hammering small nails, tacks and panel pins into wood. The round end is used to shape metal.
- **Claw hammer** heavier than a ball pein hammer. This is a common hammer. It has one flat side and one clawed. The clawed side can be used to pull nails out of wood. The flat side is suitable for hammering in larger nails and for flattening wire.
- **Lump hammer** a very heavy hammer with a large double-faced head. It is suitable for flattening lead, breaking concrete to make fragments, and breaking wood battens into shorter, irregular lengths.

Screwdrivers
Most people already have a variety of different sized screwdrivers. The two most common types are:

- **Flat head** for screws with a single slot on the top.
- **Phillips screwdriver** for screws with a cross head. This screwdriver has a positive grip on the screw making it much easier to use.

Blowtorch
You may already have a heat gun or a small blowtorch for cooking, but neither of these is powerful enough to create a blackened shou-sugi-ban surface on wood.

Blowtorches are readily available in DIY shops. They have two parts, the top section attaches to the gas canister and controls the flow of gas. Canisters can be easily replaced when empty. The gas canisters I use are a butane/propane mix but the retailer will be able to advise on the most suitable gas canisters for the blowtorch you are purchasing.

Some come with a plastic stand to fit to the bottom of the canister which prevents it accidentally falling over and causing a fire.

ABOVE Construction preparation using electric drill and table saw

I first used a blowtorch when experimenting with copper sheet and tin and was amazed at the way the application of heat affected their colour. I then moved on to burning wood to darken it and was hooked.

This tool must be used outside and away from anything flammable. The wind often blows the flame out, so one that lights automatically with the click of a button is easier to use than one that has to be lit manually.

A heavy-duty blowtorch is best if you intend on doing a lot of burning. I would advise that you not only follow the manufacturer's instructions but also research potential safety issues. I always wear thick gloves when holding and turning the wood.

Multi-purpose heavy-duty scissors

These are one of the most useful tools I have. They cut wire, wire mesh, metal sheet such as copper, lead and tin, and thick cardboard. They are the first tool I try when cutting material I haven't used previously.

Secateurs

These are ideal for cutting natural materials such as willow, sticks and twigs.

Pliers

There are many kinds of pliers but those I have found most helpful are ones you may already own:

- **Needle-nosed pliers** with narrow jaws that are thinner towards the tip.

- **Combination pliers** with wider flat jaws. These can also be used to cut wire.

Pliers can be used for pulling needles through thick or layered materials, twisting wire, tightening and loosening nuts, holding panel pins in position while hammering, and many other activities you will come across when constructing work.

Hack saw

A hack saw has a U-shaped frame with a handle at one end. It has a thin, small-toothed blade attached to the frame with pins at each end, allowing the blades to be replaced easily if they break. It can cut through most materials but is best used to make short distance cuts. For example, I use a hack saw to cut through steel reinforcing bars and narrow widths of wood. It is a slow process that takes a lot of effort, but it will also cut aluminium, copper, brass, iron, plastic and more. Electric saws are preferable for deeper cuts.

Gimlet

This is similar to an awl for making holes which many stitchers have in their sewing kit. Gimlets are slightly different as they have a screw end to drill small holes in wood without it splitting. It is a very handy tool for starting a hole in preparation for using screws.

Drills and drilling

An important requisite when working with hard, resistant materials is the ability to make holes to aid the construction process. Electric drills are essential for this. You will also need a variety of different sized drill bits. You can buy sets of different sized drill bits suitable for using with wood, metal and masonry.

Handheld electric drill

This needs to be powerful enough for the materials you choose to work with. My first plug-in corded electric drill could drill through wood but not slate. I tried a cordless drill but this also didn't have enough power. I now have a large plug-in corded electric handheld drill that is heavy to hold but much more powerful and will drill through anything.

Some electric drills also function as electric screwdrivers. These come with interchangeable screwdriver and drill bits.

Small cordless rechargeable electric screwdrivers are excellent tools if you plan on using a lot of screws as they allow the user to screw and unscrew as necessary. They usually come with a selection of flat head and Phillips screwdriver bits. Not only do they speed up the process of screwing but they save your hands from blistering.

Electric mini multi-tool

My favourite and most useful drill is part of a Dremel multi-tool (other makes are available). It is a very small electric handheld tool that comes with tiny drill bits:

- smallest: 0.79mm (⅟₃₂in)
- largest: 3.18mm (⅛in)

It is ideal for making small holes in delicate materials. I used this to drill holes in the top of the bark cloth pupae in *On the Brink* (page 68) so that wire could attach them to the sticks, and also to drill into fragile bark to enable the neoprene cord to be threaded through in the *Habitat* series (page 91). It has many small attachments for cutting, grinding, sanding, etc.

Dremel workstation

An invaluable general-purpose accessory is the Dremel workstation, which acts as a specialist drill stand. It holds the drill, changing the handheld drill into a mini pillar drill. A pillar drill has a handle to lower and raise the drill bit, allowing accurate placement of holes.

Because this was so useful, I also bought a full-size pillar drill for drilling into thick media. This has the advantage over a handheld drill of holding the material in position while drilling. While it is a very practical tool for wood, it is equally useful for drilling through 25mm-wide (1in) neoprene cord. By clamping it in position I could drill through the cord without it twisting or the bit slipping off the curved surface. I used this on *Boundary Lines* (pages 12 and 98).

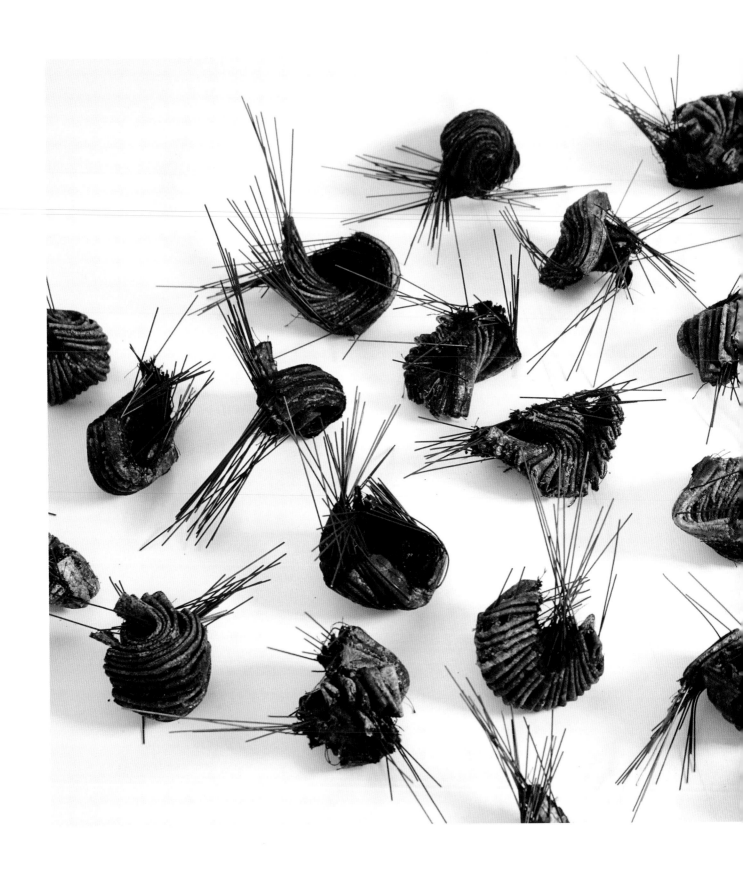

ABOVE *Extinction*: cotton fabric, wax, wire; installation consisting of 40 small sculptures; an example of working in multiples

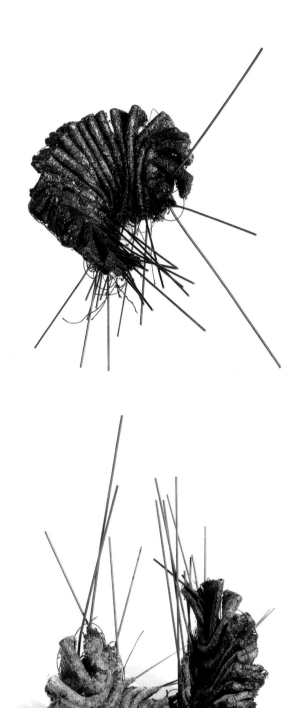

Saws and sawing

Sawing with non-electric hand saws is slow and requires a lot of effort unless cutting through thin pieces of wood, so I find it preferable to have an electric saw.

Electric saw – handheld

- Advantage: It speeds up the process. The model I bought came with interchangeable blades for metal and wood.
- Disadvantages: It is extremely noisy, so ear defenders are necessary. It vibrates quite strongly when switched on. It can be difficult to cut long, straight lines with freehand sawing.

Electric saw – table saw

In order to cut long, straight strips I ended up buying a table saw. It was the perfect tool when I was making *Boundary Lines*, *Limitations 1* (page 66) and *Blocked Off* (page 42) as it made preparation much faster and consistently produced identical shaped pieces.

A table saw has a circular saw mounted under a metal table surface, with part of the blade protruding above the table. This cuts continuously when switched on, allowing a more controlled method of sawing. It has measurements marked on the metal table and a metal rip fence (bar) to keep the wood in position for accuracy.

My table saw has metal legs to lift it to a working height. It came with a plastic push stick to move the wood forward and keep the hand away from the revolving blade.

Protective kit has to be worn: thick gloves, safety glasses, ear defenders, face mask to prevent breathing in sawdust, and hair tied back.

Cutting slate with a tile cutter

To cut slate, I use an electric tile cutter. This is used outside. It is similar to a table saw in design in that it has a circular saw set under a metal table marked with measurements and a movable rip fence (bar) holds the slate against the blade for accurate cuts; However, while part of the saw protrudes, the remainder of the saw sits in a container of water which cools the blade as it is revolving making the cutting easier. Safety kit is required: waterproof gloves, face mask, safety glasses, and hair tied back.

Safety

The potential risks are wide-ranging (use of tools, potentially dangerous materials, trip and fall hazards, electricity use near water) and are compounded by working alone. I mitigate these risks by assessing the potential problems before each stage of work. I take the appropriate precautions – clearing the working space, using protective kit, ensuring good ventilation, and operating tools in the recommended manner.

Display considerations

From experience, I have learnt that it's easy to get carried away making a piece only to find out when it's completed that you haven't considered how to display it. One of the first things I now do when planning a work is to decide if it should hang on the wall, be suspended or be free standing. I might change my mind as the piece progresses but if the intention is to hang it in some way, it is a good idea to think ahead how to incorporate a hanging device.

The usual methods of wall hanging for framed work are not suitable for sculptural work, especially if constructing with paper or other delicate materials. If the piece is lightweight, I often thread small loops onto the back, made either from wire or from the same material as the construction, so it can be hung from a screw in the wall. Another option is to make the hanging device a way to convey meaning like the entomology pins in *On the Brink* (page 68).

Element construction

Most of my work features repetition of elements in one form or another. The elements may be similar to each other or be deliberately varied in some respect. It is the method I use to build most of my work. My constructions are brought together in several steps. Much preparation is required in advance before assembling the final form. In this section, I will discuss the pre-construction preparation that went into producing several of my pieces. The first step, of course, is collecting the individual materials that will make up the whole.

PREPARATION CASE STUDY
Boundary Lines

As mentioned previously (page 12), this piece was inspired by the theme of boundaries. I wanted a variety of identically sized strips of wood and slate to imply control, to define boundaries and to provide three-dimensional structure. All the hard materials chosen are those used in fencing, referencing actual boundary markers in the landscape.

Materials wood, slate, metal mesh, willow, scrim, silk noil fibres, neoprene cord in three different thicknesses, paper yarn, nuts, bolts, screws, mending plates, liming paste

Techniques used in the preparation of the separate elements included:
* Sawing and drilling wood to the desired length and width
  Tools table saw, pillar drill
* Cutting and drilling slate to size
  Tools slate cutter, handheld electric drill
* Cutting willow in short lengths
  Tools secateurs
* Cutting metal mesh to size
  Tools multi-purpose heavy-duty scissors
* Making strips of felted paper

- Knotting the neoprene cord to the short willow sticks
- Fastening the paper yarn to the metal mesh
- Rubbing liming paste on the wood to blend it with the other elements and protect it

PREPARATION CASE STUDY
Extinction

Extinction (pages 78 and 79) is an installation of forty small constructions. Each of these constructions is an element in the installation, and each construction is made up from even smaller repeated elements. It was inspired by exploring the theme 'human impact on the environment' and hints at the potential mass extinction of species. The colour, surface treatment and straight wires were chosen to imply the small life forms' possible demise. The following description is the preparation of the small elements before they are combined together to make up the constructions.

Materials black cotton fabric, wax, wire, thread
Techniques used in the preparation of elements:
- Covering the fabric with hot wax
  Tools electric wax pot, old paintbrush
- Leaving waxed fabric to cool and harden
- Crumpling and crushing the fabric to crack the wax, rather like batik but the wax is left on the fabric
- Cutting the fabric into strips
- Folding the fabric in half with wire inside, ends protruding and stitching it closed
  Tools needles, sewing machine

Repetition is an important aspect. This process was repeated many times until I had enough strips to form the small constructions. I started hand stitching the strips: this was a very slow process due to the thickness of the waxed fabric, so I decided to machine stitch them instead. This was also problematic as it was a struggle to feed the thickness of the waxed fabric through the machine. After a lot of thought, I solved the problem by wrapping each strip in tracing paper so I could still see the edge of the fabric to stitch, and the wax didn't clog up the machine. The paper was easy to remove: it had been perforated by the sewing machine so it just tore off.

PREPARATION CASE STUDY
Ecotype

This is another piece that I developed from investigating the theme 'human impact on the environment'.

Materials mitsumata paper (from Himalayan foothills and made from the inner bark or bast fibre of native plant mitsumata), lokta paper (from Nepal and made of the fibrous inner bark of mountain shrubs) – both are sustainable and renewable resources, A4 printer paper, paper yarn, tea dye, wax, gesso

ABOVE Detail of *Ecotype*

BELOW Preparing elements for *Ecotype*

Techniques used to prepare the elements:
- Printing words and phrases from my research on the printer paper
  Tools computer, printer
- Texturing papers with wax, gesso and tea dye
  Tools wax pot
- Tearing paper
- Manipulating torn pieces into desired shape

Repetition is evident again. All the paper elements are the same shape but vary in size, type of paper and surface finish. A large quantity was made before being assembled into the final form.

PREPARATION CASE STUDY
Hanging by a Thread series

Five constructions make up the *Hanging by a Thread* series (page 49). They were inspired by a brief for an exhibition called 'Package Tour'. I decided to use packaging materials to reference the exhibition title and realised it also fitted in with my theme of 'human impact on the environment', as trees had to be cut down to make the paper and the corrugated card. The burning of the torn elements was informed by the forest-clearing technique of slash and burn.

Each of the constructions was formed from a different selection of the materials listed here.

Materials brown paper packaging, corrugated cardboard, cartridge paper, bristles, willow, linen thread
Techniques
- Tearing paper and card into small shapes
- Burning the edges of all the shapes
  Tools tweezers, candle
- Sealing each piece with diluted PVA
  Tool paintbrush

A mass of torn and burnt elements were prepared before assembling. I am never sure how many elements I will need. Sometimes I have to make more if I run out before I feel the piece is resolved.

RIGHT Preparation of the burnt corrugated card elements for one of the *Hanging by a Thread* series

6

Constructing a three-dimensional form

LEFT *Biform* detail; broken fence panels, willow, neoprene cord, nuts and bolts, fabric covered wire

Constructing a three-dimensional form

Two objects' incompatibility is not an excuse not to use them

OLU AMODA

A constructed three-dimensional form is built up from separate elements. Any material or found object can be incorporated into the structure. Finding a solution to join contrasting materials together is like solving a puzzle and is part of the fun of constructing.

Assembling the elements

After preparing or collecting the elements, my next stage is finding ways to assemble them. Below are some of the textile and non-textile joining techniques I have used.

- Stitching
- Twining
- Weaving
- Threading
- Wrapping
- Knotting
- Layering
- Tying
- Folding
- Bending
- Pinning
- Casting
- Screwing
- Trapping
- Bolting
- Clamping
- Bonding

Using traditional textile techniques with hard materials or found objects is often a challenge, but I find it valuable as a way of introducing meaning into the work, by evoking unexpected associations and connecting it to wider issues.

Joining hard materials

When assembling wood or metal elements, the first thing I do is work out how and where to make holes and what I will use to hold the elements together. I have a selection of drill bits of all sizes. At first, I used to choose the drill bit to match the width of the thread, but discovered that threading through the holes is less of a struggle if the hole is a fraction larger, especially if the thread is going through several layers of material. For example, I use a 3.5mm (⅛in) drill bit to make a hole for 3mm neoprene cord instead of a 3mm hole that is the actual size of the cord. I also check the edge of the hole; if rough or sharp, it may over time rub against a soft thread and break it, so I use wire instead.

If I want to drill a large hole in wood, I make a pilot hole first using a small drill bit and then use a larger drill bit of the size required. I find it easier to widen a hole than start a large hole from scratch. The pilot hole also keeps the bit from moving out of position when the drill first touches the wood.

To drill a hole in a smooth shiny material like metal I use a hammer and nail to tap a dent in the position for the hole first before drilling. This prevents the drill bit from slipping out of place, making scratch marks on the surface or making the hole in the wrong position.

Once the holes are drilled, it is possible to poke stiff thread alternatives or needles straight through the hole before fastening the elements together by stitching, tying or knotting.

If I need to use a needle, I consider the combined thickness of the elements. Most needles I use have large eyes to accommodate paper yarn; some have sharp points and others are blunt. I find bodkins

RIGHT Examples of stitching into wood and slate with paper yarn and wire

88

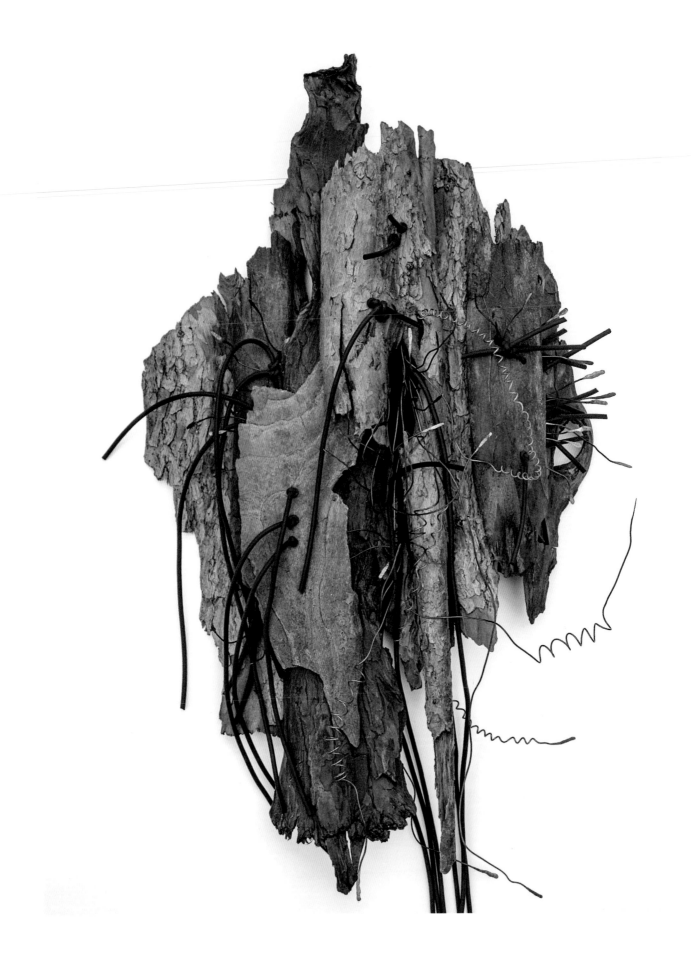

the most useful for my style of working. Whichever needle I use, I usually use pliers to pull it through if it gets stuck, which happens frequently.

Giving structure

As an artist focusing on 3D construction there is a reason why I work with particular joining techniques for specific pieces: I find it is vital to be able to bring the appropriate types of structural elements into the mix to enable the realisation of the construction.

I frequently use rubber, wire or paper yarn as alternatives to thread to hold the elements together, but the option I select will depend on the materials being assembled and the stability required. Apart from structural considerations, there are also aesthetic reasons and conveying meaning to think about.

Visually, I consider whether I want the joining process to be the same throughout the construction or be combined with contrasting methods. In some instances, I think about (or experiment to decide) whether the construction method should be hidden depending on whether or not it is being used to convey meaning. It is important to me that the joining process is integral to the construction and carries the concept or narrative through to conclusion.

Another aspect to think about is whether I have the tools I require to assemble the elements. Techniques such as weaving, twining and wrapping need no tools other than scissors, possibly wire cutters if using wire, or secateurs if assembling natural materials.

The type of materials being put together determines which tools to use.

Below are examples showing how I use textile techniques functionally as structural elements, in addition to them being integral to the composition.

CONSTRUCTION CASE STUDY
Habitat series

The *Habitat* series consists of four separate bark constructions that were developed from research for the theme exploring human impact on the environment. Using the yew bark for my first construction in the series inspired me to collect other types of bark, as I enjoyed using the material and working out the best way to assemble the fragments. The constructions could have been exhibited as multiples as parts of one piece, but they were conceived as separate pieces and I felt they were visually strong enough to stand alone as a series.

Idea to create new habitats from fragments left over from habitat destruction
Materials bark, neoprene cord, copper wire
Techniques drilling, threading, stitching, knotting
Tools small electric drill, heavy-duty scissors

The lightweight fragments of bark are held together by manmade materials, copper wire and neoprene cord, suggesting that what man destroys he can also restore. After drilling holes in the bark, it was possible

LEFT *Habitat* series: knotted neoprene cord and copper wire hold the bark fragments together (W30 x H62 x D10 cm / W12 x H24½ x D4 in)

LEFT AND ABOVE *Adrift*; willow, wire; soumak weaving (W60 x H150 x D38 cm / W23½ x H59 x D15 in)

Using metal fixings

My pieces on the theme of boundaries combined wood and slate with textile elements. The hard, heavy materials required strong methods of joining the elements together, so I used metal fixings functionally to give structural stability to the constructions but also to convey notions of control and restriction.

Metal fixings include nuts, bolts, screws, washers, threaded rods, mending plates with pre-drilled holes, and wire mesh, all purchased from hardware stores. They all come in different sizes and widths and are best used when attaching directly in an accommodating material such as wood. A variety of screwdrivers are useful for the many different types of screws available. If using a large quantity of screws, an electric screwdriver is invaluable and less strain on the hands. It is also useful for removing screws.

to stitch, thread and knot these thread alternatives, continuing until the form felt structurally stable.

CONSTRUCTION CASE STUDY
Adrift

This is a second piece developed from researching the theme for the exhibition 'Ebb and Flow' in Grimsby where I explored the decline of the fishing industry.

Idea the discarding of traditional fishing gear, a metaphor for the dying industry
Materials willow (withies), rusty iron wire
Tools wire cutter, secateurs

The withies were chosen to reference traditional lobster pots and are held together with wire by the soumak weaving technique. The weaving is deliberately loose as I wanted the structural instability to evoke a feeling of falling apart.

CONSTRUCTION CASE STUDY
Field of Pollution

This piece was made for a small format textile exhibition where the criteria for 3D work was that it should be no larger than 20cm (8in) in any direction.

Theme air
Idea to explore how the human activity of burning fossil fuels causes air pollution, which produces acid rain. This reduces soil fertility, prevents photosynthesis, stunts plant growth, and kills trees
Materials lead, paper yarn, tea dye, calico
Tools lump hammer, multi-purpose heavy-duty scissors, sharp needle, pliers, small drill

The piece represents an area of land contaminated by acid rain. The lead was selected for its connotations of toxicity. It was hammered to make it more malleable,

RIGHT *Field of Pollution*: lead, paper yarn and calico. Lead elements appliqued with paper yarn to calico on wood base (W12 x H15 x D6 cm / W4¾ x H6 x D2¼ in)

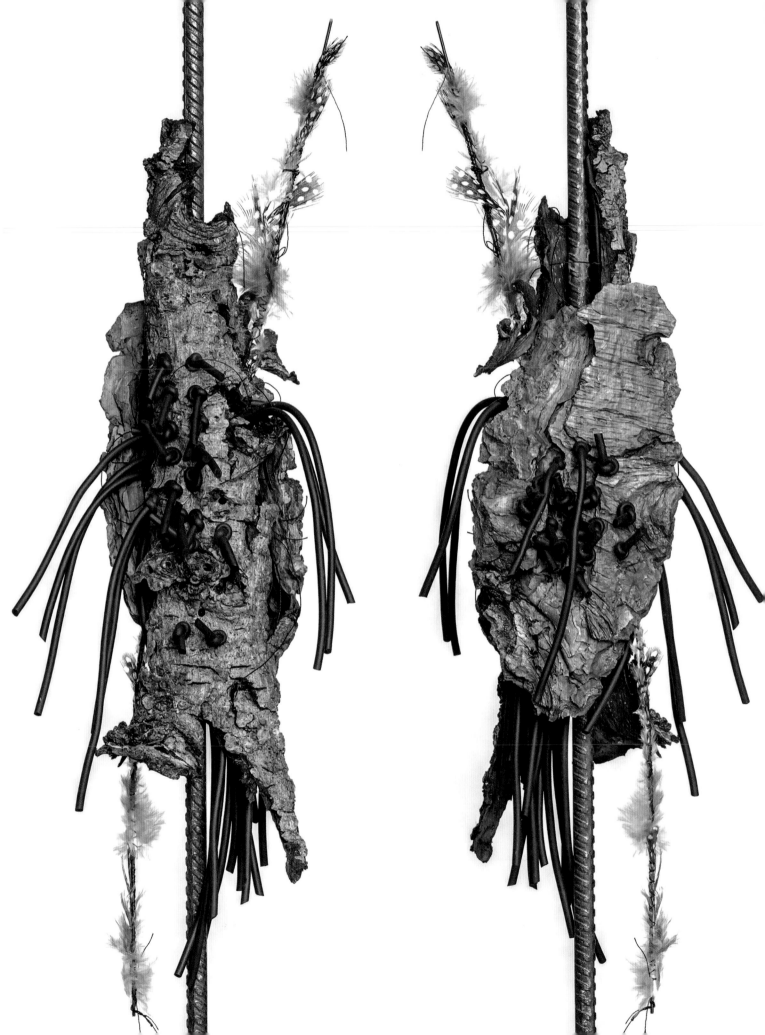

LEFT *At Stake* (detail); three of seven stakes, bark, neoprene cord, linen thread, feathers, rebar; stitched & knotted (W20 x H100 x D10 cm / W8 x H39¼ x D4 in)

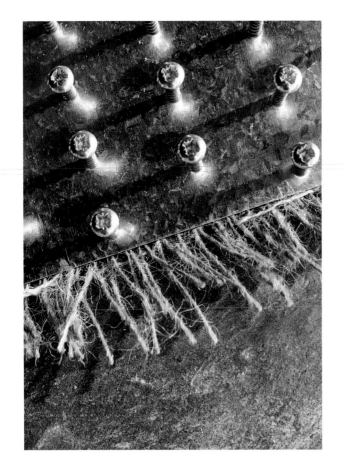

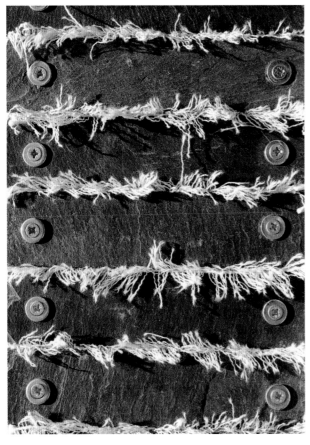

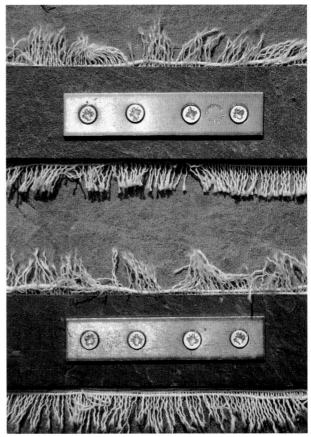

then either cut or torn into small pieces, which were hammered again and manipulated to make them three-dimensional. Two small holes were drilled in each piece to allow them to be appliquéd to the base.

The paper yarn was tea-dyed, the colour suggesting dying vegetation. It was then used functionally to attach the lead to a thick calico base with knots to hold them in place.

The exhibition was overseas, so the small format of the piece allowed the use of lead, which would not have been practical on a larger scale due to its weight.

CONSTRUCTION CASE STUDY
At Stake

This piece was developed from the research for my self-directed theme 'human impact on the environment'.

Idea to use relevant materials to highlight the damage being done to nature by the construction industry
Materials rebar, bark, wire, feathers, saw blade, neoprene cord, linen thread
Techniques machine stitch, knotting, hand stitch
Tools small drill, sewing machine, needle

The installation consists of seven suspended rebar stakes. The stakes go through the centre of the bark structure to reference the construction industry breaking up the forest habitat. The feathered wires

protrude from each unit to hint at birds escaping the destruction. Each bar has the bark and other elements attached to it with knotted neoprene rubber and linen thread. The structural stability of the pieces comes from the strength of the rebar combined with the tight knotting and stitching.

CONSTRUCTION CASE STUDY
Boundary Lines

As previously mentioned (page 12), this piece developed from research for the boundaries theme. Below are the joining techniques using metal fixings which were required to combine the disparate resistant materials.

Materials wood blocks, slate blocks, screws, large and small washers, hexagonal bolts, square nuts, mending plates, felted silk noil paper, neoprene cord, paper yarn, scrim
Tools Phillips screwdriver (cross head), pliers

All the elements are attached to seven base planks. The slate strips and wood blocks have been arranged into areas with defined boundaries. They are fastened to the base planks with nuts, bolts, screws and mending plates, trapping the textile elements but not quite holding them in place. Drilling holes completely through wood makes it possible to assemble the elements with bolts, nuts and washers.

A different solution was needed when assembling layered pieces like *Limitations 1* (page 66) and *Blocked*

LEFT Examples of metal fixings holding down fabric; large mending plate with screws in predrilled holes, metal mesh, screws and washers, small mending plates with screws

Off (page 42). Drilling holes in each layer made it possible to push threaded rods the full height of the construction, which was determined by the 50cm-long (20in) threaded rod. Nuts were screwed onto the rod and tightened at both ends.

CONSTRUCTION CASE STUDY
Under Concrete

This was inspired by the research for my self-directed theme 'human impact on the environment'.

Idea biodiversity is under constant threat from developers concreting over the landscape and destroying habitats
Materials cotton noil fibres, bristles, wire, concrete
Techniques cotton paper making, assembling, concrete casting

This is an installation of 70 small concrete units, each embedded with a silk paper and bristle form. The bristly elements allude to seeds or other small life forms. The separate concrete pieces reference the destruction of habitats. Irregular shaped moulds were made out of corrugated card and attached to the melamine base with masking tape. A fibre organism was placed in the bottom, covered in concrete, and left to cure for a few days before tearing off the card moulds.

By juxtaposing the soft, delicate cotton paper with the hard, heavy, unforgiving concrete, the intrinsic qualities of both materials emphasise the other's qualities. Casting the small life forms in concrete conveys their vulnerability and the potential consequences for biodiversity.

CONSTRUCTION CASE STUDY
Secret Support (extra firm hold)

The theme for this piece came from an exhibition. It was an opportunity to respond to items in Platt Hall Costume Museum, Manchester. Inspiration came from close examination of the eighteenth-century stays and stomachers in the women's underwear collection. After researching their construction and the placing of the boning concealed within stitched channels, I decided to play on the qualities required for foundation garments by using materials used in the construction industry.

Idea to use materials from the construction industry as a metaphor for the function of the shape wear
Materials concrete, reinforcing bars, wire, industrial staples, metal mesh, screws, fabric
Techniques mould making, concrete casting, embedding
Tools hack saw, electric drill with concrete mixer attachment, wire cutters

Using a simplified pattern of a stay, I made my own moulds from four pieces of wood, held together with masking tape, and attached them in position

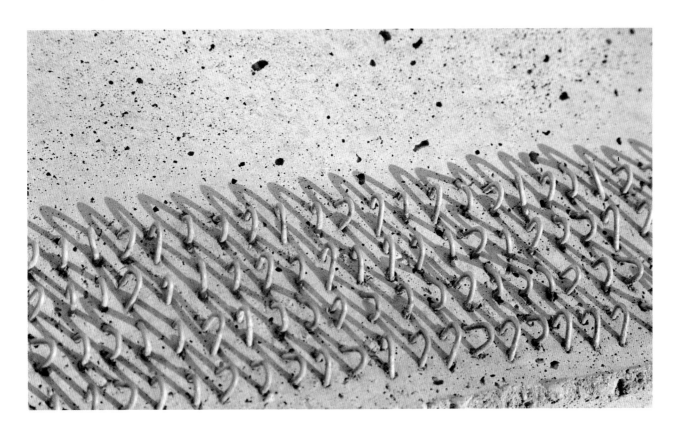

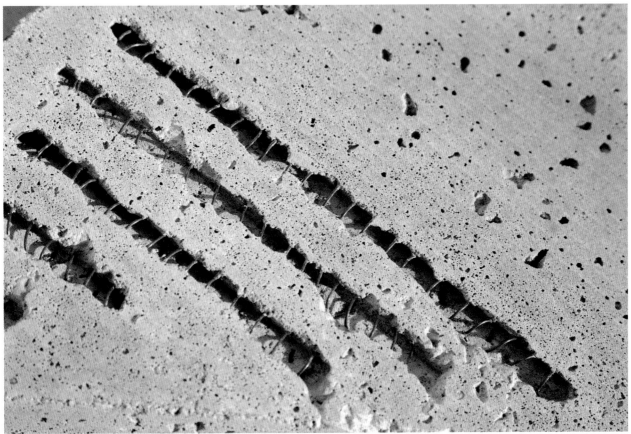

ABOVE Industrial staples in concrete, embedded wire stitches in concrete; both reference the surface stitching and hidden stitching on stays and stomachers

ABOVE *Under Concrete* (detail); felted cotton paper forms
embedded in cast concrete; installation of 70 units

to bases made from old melamine shelves left over from a dismantled kitchen. While experimenting, I discovered the concrete could be removed fairly easily from this.

The sides of the mould were covered in cling film to prevent the concrete sticking to the wood. The prepared metal elements were placed on the flat base inside the mould, replicating the pattern of the boning. After mixing the concrete, I poured it over the metal pieces to the top of the mould. From earlier sampling I knew that the top surface in the mould needed to be finished properly otherwise it looked unsightly, so each mould was covered in hessian touching the concrete, and a piece of cardboard was pressed on top to keep it in place. The concrete, in combination with the other materials, was chosen to signify strength and control.

It took several days for the concrete to cure before I could remove the mould. This was easy to do, as each piece of wood peeled off separately having only been held together with the masking tape. It was then a case of checking if the concrete was hard enough to turn the shape over to see the result.

This was my first use of concrete in a piece. The images (page 101) show details of two of the eight units from the installation:

- One has hidden wire stitches embedded in the concrete stay.
- The other has industrial staples to reference the surface stitching.

Constructing multiples

Eva Hesse was a German-born American artist making sculptures from latex, fibreglass and plastic. She often constructed her pieces from multiple modular units such as in *Repetition 19* and *Accretion*. This approach has been a major influence on my making over the years. Other major artists working in this way include:

Carl Andre *25 Blocks and Stones; Flanders Field*
Magdalena Abakanowicz *Backs*
Antony Gormley *Field*
Michael Brennand-Wood *Humpty Dumpty; Dreams Within the Here and Now*

A lot of my work is fairly small-scale, and when I have tried to scale up it doesn't always work. This is partly due to the materials I use. Therefore, exhibiting many small, related elements together enables me to produce a larger work. When doing this, the separate elements become part of something larger, taking on extra layers of meaning and presence due to being part of a group.

On a practical note, this also makes the piece more adaptable to display, allowing it to change to fit the space available. It is also a good strategy to use if working with heavy materials. This is the approach I employed when making *Under Concrete* (page 100), where the large number of small fragments were packed in manageable batches in separate containers, making it easier for carrying and transporting.

RIGHT *Endangered* series (2 of 3 pieces); lokta paper, bristles, wire

At Stake (page 99) is an example of several separate elements exhibited as one piece. Each of the rebar stakes is different but they are linked together as one piece due to proximity and the combination of materials.

Series

Usually when making a piece, I find it sparks ideas for related pieces. By acting on these ideas, I build up a series of similar, but separate, pieces all exploring the same aspect. Sometimes, I realise I have so much material left over that I carry on making variations of the original construction from it. The following are some examples of series I have made.

CONSTRUCTION CASE STUDY
Endangered series

Research from the theme of human impact on the environment inspired the three pieces in this series. They represent the millions of small life forms, many of which are as yet unknown to science, that are in danger of becoming extinct through loss of habitat caused by human activity.

Materials lokta paper, bristles, paper-covered wire
Techniques tearing, burning, assembling, threading

The tearing and burning of the elements in *Endangered* was influenced by the slash-and-burn agricultural farming method used in forest clearing.

The method of construction was informed by the repetition and tessellation observed in the electron microscope images of seeds mentioned previously.

Each piece was constructed from a mass of torn paper elements. The elements were a different shape for each individual structure. Each paper element is burnt at the edges using a candle. I used tweezers to hold and turn the small pieces in the flame. To prevent the burnt areas breaking off, both sides of the paper shapes were sealed with diluted PVA. When dry, they were bonded together in pairs with bristles trapped between them. The ends of the bristles protruding were also singed.

After this preparation, the constructions could be assembled by threading the elements onto wire.

The shape and repetition of the paper elements determined the differing form of the finished pieces. The bristles were strong enough to hold the weight of the paper and raise the structure above the surface they stand on.

I kept the studio door open for ventilation when using the candle indoors and I wore a face mask. The procedure was carried out on a non-flammable surface away from draughts and any flammable items. I also had a damp cloth next to me to immediately extinguish any smouldering.

The *Endangered* series was made for a miniature paper exhibition. I later used this method of construction (without the bristles) on a much larger scale when making *Ecotype* (pages 70 and 83).

RIGHT *Hanging by a Thread* series – construction details of two of the pieces from this series

Hanging by a Thread series

I initially made two of these constructions, which were exhibited and sold as a pair. I decided to make more for myself as I was pleased with how they had turned out. The structures formed suggest natural forms like seed or pods. I made three more variations that are similar but not the same as the originals.

Idea response to an exhibition title called 'Package Tour', to explore how the manufacture of packaging material affects nature
Techniques twining, piercing, tying, threading

The preparation of the elements was described previously (page 84). To assemble them, I twined, wrapped and knotted the bottom ends of the willow structure with neoprene cord to hold them together. I then threaded the paper elements onto each willow rod, alternating them with bristles to keep the paper in place. The paper pieces on some were sparse, leaving the willow showing to connote withering away. The construction was completed by fastening the top rods together with more neoprene cord. The pieces are displayed suspended by a black linen thread. The materials and processes used, and the overall form and method of presentation were informed by the idea.

Hostile Landscape

This is a concrete work on a larger scale than my previous concrete pieces. It is an installation comprising nine separate units. It explores the idea of concreting over the landscape but interprets it differently to the *Under Concrete* (page 100) installation.

Theme human impact on the environment
Idea to hint at the progressively diminishing habitats created by paving over land in areas including parks and gardens
Materials concrete, cotton noils, bristles
Techniques casting concrete, felted cotton paper making

I prepared a lot of felted cotton fibre paper elements, incorporating bristles in some. These were then assembled into constructions representing different species of seed, inspired by the electron microscope images of seeds mentioned earlier (page 31). A mass of small seed forms was made.

With this being a bigger, more ambitious piece, I mixed a larger quantity of concrete using an electric drill with a concrete mixing attachment and a heavy-duty builder's bucket.

Some of the small forms were sprinkled over the base of hand built moulds before being covered in concrete. It was then left to cure for a few days.

For this piece, the presentation of the units alludes to paving slabs. Each one is 40 x 40 x 3cm (16 x 16 x 1¼in). Some of the slabs have the fully formed cotton fibre seeds sprinkled onto the bare concrete, implying that they are falling on stony ground. Other cotton fibre seeds have been embedded in the concrete, some to the point of disappearing. The juxtaposition between the soft felted forms and the hard concrete aims to emphasise the vulnerability of nature.

Dismantling, recycling and reusing

As a lot of my work explores environmental issues, I am keen on recycling and try to be mindful of the materials I use. Nowadays it is more important than ever to recycle whatever we can to help the planet. I find it difficult to reconcile using concrete in my work as I know it is environmentally unfriendly, but I try to mitigate that by the fact that by using it I am highlighting that exact point.

I am not reticent about dismantling old pieces of work if they are too big to store or are ones I no longer like. I then reuse the materials in other pieces.

The wood used in most of my boundary-themed pieces are reclaimed planks that were once the floor of a church hall. I reused them to make a large installation that controlled access to the gallery space. After dismantling the installation, the wood has been reused in pieces that explore access and control.

Another large boundary-inspired piece that I dismantled was made from wood I had collected from beaches – a mixture of driftwood, old fence posts, broken window frames and furniture, all nicely weathered by the sea. I have since reused many of these bits in other work.

I sadly had to dismantle my large paper sculpture, *Ecotype* (pages 70 and 83) because it was damaged. However, I kept the paper elements and recycled them as elements in a new sculpture, *Mutation* (page 20).

Marie-José Gustave

Marie-José Gustave is an artist of French Caribbean heritage. She has been creating works with paper for about fifteen years. A self-taught artist, her work is based on experimenting with materials and combining different craft techniques, including basket making, knitting and crocheting.

Originally from Guadeloupe, born in France and living in Quebec, Canada for more than twenty years, Marie-José creates installations and large pieces of works on the subjects of interbreeding, migration and the search for identity.

From her training in clothing production, Marie-José has kept a taste for form and volume. She uses the flexibility and rigidity of paper yarn to create volumes, and the stitches of knitting, crochet and braiding to play with shadows and light. Those shadows evoke preconceived ideas and how our gaze can change.

Paper thread illustrates the flexibility necessary to adapt to a new society, and the universal ancestral techniques that Marie-José uses speak of transmission.

In her recent explorations, she works on the interaction of clay and paper thread with the technique of basketry. The paper thread and the clay constrain, oppose, bend and adapt to each other in the braiding process that binds them together. The result is hybrid constructions with a singular aesthetic.

Marie-José comments: 'Since 2018, I have been combining clay with my work with paper thread. At first I made grids of woven and knitted paper thread on top of which I stitched handmade clay forms. Since 2020 I've been working on a series of sculptures with round clay forms woven to one another, using basketry twining techniques.'

LEFT *Root.* Clay, paper thread (H35 x D27 cm / H13¾ x D10½ in)

Kieta Jackson

Kieta Jackson is a textile artist who has established a distinctive area of practice by using textile techniques in metal. Her small-scale sculptures have a sense of antiquity, from another place and time. She works primarily with wires and sheet metal, and builds vessels and forms of crocheted and woven fabric.

Kieta explains: 'The process starts with a fine-gauge wire, which is then woven on a loom by hand, creating a malleable fabric that is manipulated into the sculptural identity. I am always striving to create a harmonious relationship with material and forms, so the sculptures resonate found artefacts, archaeological remains and elements from tribal artefacts.'

While her pieces adopt shapes borrowed from unearthed objects, she also distresses and creates a patina on the surfaces of the metal, conveying deterioration and corrosion.

LEFT *Caballarius 1*. Wire, metal sheet (W10 x H37 x D9 cm / W4 x H14½ x D3½ in)

Nerissa Cargill Thompson

Nerissa Cargill Thompson is a designer, maker and facilitator with over twenty years' experience of professional and community practice. She originally trained in Theatre Design but through her community arts practice, her interest in fibre art and desire to develop personal artwork grew. Her practice currently explores environmental themes through three-dimensional textiles and photography. She has developed a unique way of highlighting plastic pollution by using litter as moulds in which to cast her cement and embroidered textile pieces. Nerissa comments:

'My work investigates change over time, not just eroding or decaying but new layers of growth, creating juxtapositions of structure and colour. Recent work highlights the issue of plastic pollution and the permanence of disposables, through sculptures that combine embellished textiles and cement cast in plastic waste. These sculptures invite us to consider the packaging that we use and discard on a daily basis – objects that are so lightweight and seem so insignificant that we barely notice them. Casting them in concrete gives this waste a greater physical and psychological presence that mirrors their true legacy and the seriousness of the ecological catastrophe to come. Naturally-inspired textures and embroidery emphasise the way our waste becomes subsumed into the natural world, giving a distinct contrast to the manmade structure of the packaging.

I often work in series or multiples. As I use actual litter and household plastic waste to cast in, the size of each individual piece is predetermined. Using multiples allows for the development of larger artworks but also creates a sense of accumulation. The dismissal of plastic waste as "just a bottle" or "only a mask" is part of the problem; it is the build-up that has tipped us into crisis. Even when working in multiples, each one is cast from a unique piece of litter rather than a mould producing duplicates. Each one counts.'

TOP LEFT Message in a bottle series, *Three Green Bottles*
TOP RIGHT *Straw Problem*
BOTTOM *More than Jellyfish*. Cast cement and embroidered textile

Hilary Bower

Hilary Bower studied embroidery at Birmingham Polytechnic in the early 1980s. She has since gone on to establish herself as a visual artist, mentor and tutor. Her work has become increasingly sculptural, with a fine art basis and makes use of cloth, metal, wood, sand and wax. The simplicity of materials, pared back form, and subtle use of different media gives her contemplative pieces an air of calm. Hilary comments on her work:

'For several years, I have been exploring and researching the notions of silence, stillness, of waiting and insignificance as things of substance; of tangibility, matter, dark and shadow, weighted and light; of occupying space. I have gathered insignificant items and objects both visually and physically while seeking to make something of greater significance, of greater resonance and presence.

Drawing and mark making underpin my arts practice. It is through this activity that I can clarify thoughts, concepts and understand what I can see in my mind's eye. To create physical presences from intangible, non-material matter is a constant endeavour within my work. It is often an intensive process to extract the clear essence of what I seem to need to make. I have to submerse myself in this state of being and working to reach the depth of focus necessary in finding the clarity from which to proceed.

Work straddles both two and three dimensions, often using multiple units to create one piece; a format I have repeated over time. In a piece titled *In Silence*, the four pouches represent a silence that is captured within. The wood that clamps the opening on either side I see as holding in that silence; noiseless but present. The simplicity of the materials employed is also important in my work with natural fibres, wood and linen amongst others being the most appropriate vehicles to translate my responses.

The use of repetition running alongside works that stand alone is ongoing, as is my need to make mixed media "drawings" that are an entity as well as a means of clarifying and questioning my creative investigations.'

LEFT *In Silence.* Muslin cotton, plywood, aluminium, nails, acrylic paint, wax (W92 x H25 x D4 cm / W96¼ x H10 x D1½ in)

Conclusion

When starting this book, my aim was to share the creative process that led me to develop my own visual language. By doing so, I hoped to show why I work with the materials I do, and how I use materials, processes and methods of construction as conveyors of meaning to engage with issues beyond techniques and the visual. I also wanted to indicate how my practice is still related to textiles despite the eclectic media I employ and the marginal position in which my work exists.

I have shown the progression through every stage of my process. The nature of my approach is quite methodical, and it gives an overview of the stages I go through in making my three-dimensional reliefs, assemblages and constructed sculptures. This gives me the structure from which I can practise creative freedom. The written, visual and material research generates many ideas and associations that creatively take me in all directions, inspiring many different interpretations. I'm always open to new ideas popping into my head, jotting them down to explore at a later stage.

Hopefully, the ideas and processes discussed will encourage some readers to explore different and new methods of working as a way of finding their own creative voice. The whole creative process is an adventure of discovery. As author Barbara Sher says: 'There's no one in the world who can do what you do, who can think and see the way you do, who can create what you can create.'

RIGHT Small assemblage; waxed wood, burnt wood, waxed paper, net, concrete, linen thread, wire

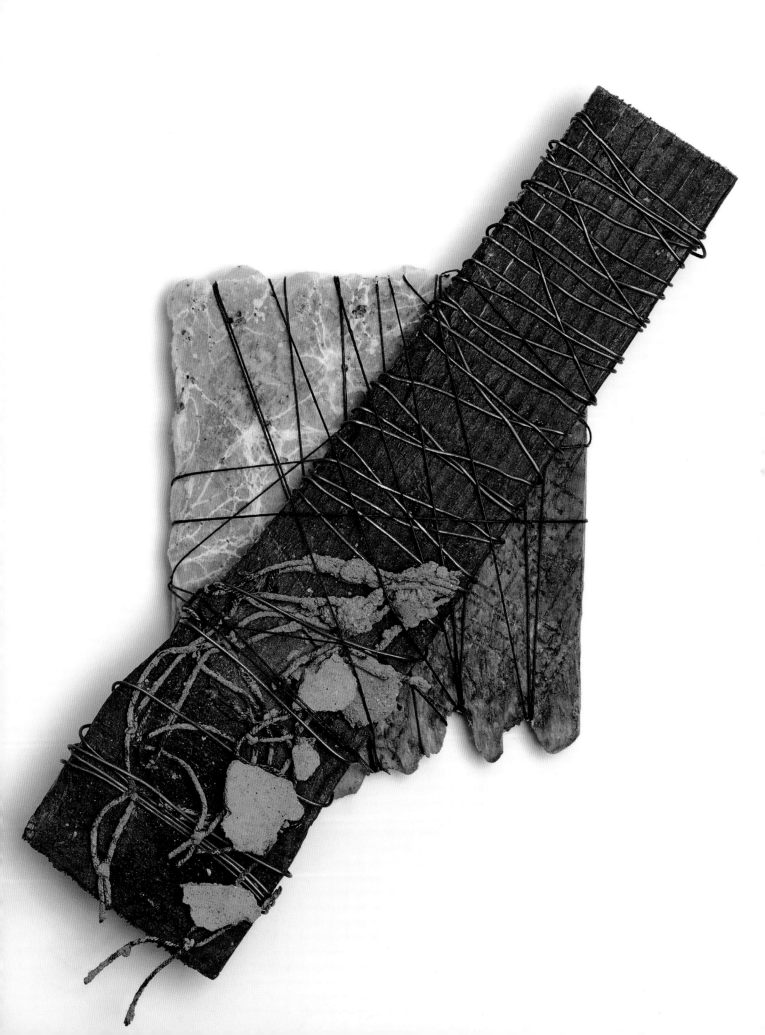

LEFT *Limitations 1* (detail)
ABOVE Details from *Under Concrete*

121

LEFT *At Stake* (detail)
RIGHT *Discarded* (detail)

Glossary

assemblage artwork made by combining diverse mixed media

barkcloth a traditional cloth made in Ugandan villages from the inner bark of Mutuba trees that are at least 8 years old. It is a sustainable and renewable material as the bark is reharvested after five years allowing the tree to regenerate. The wet inner bark is beaten with mallets creating a smooth terracotta coloured finish

City & Guilds an educational organisation offering vocational courses and qualifications in many subjects, including textiles

hybrid a mixture of previously different genre; combination of diverse media

lokta paper made from the inner bark or bast fibres of plants that are native to the Himalayan foothills. They are a renewable resource; they are cut above ground level and can be reharvested after 3-4 years

mitsumata paper see lokta paper

monofilament filament of synthetic fibre, e.g. fishing line

neoprene cord synthetic sponge rubber cord

noil short, matte fibres of cotton or silk; a by-product of combing the long fibres in the spinning process. Cotton noil is also known as comber noil

PVA a white glue (known as 'Elmer's glue in the USA) that dries clear and does not yellow. Can be diluted

rebar steel reinforcing rods used in the building industry

relief a wall-mounted artwork in which the three-dimensional elements are raised from a flat base

shou-sugi-ban a Japanese technique of burning wood

soumak a weaving technique

SWG abbreviation for standard wire gauge. Low number SWG = thick diameter wire, High number SWG = thin diameter wire

thread analogue a material that can be used as an alternative to thread, e.g. wire, fishing line

twining a basketry technique

wabi-sabi a Japanese aesthetic concept which finds beauty in simplicity, impermanence, asymmetry, imperfection and the understated

Further reading

Andrews, Oliver, *Living Materials, A Sculptor's Handbook* (University of California Press, 1988)

Art Textiles 2 (Bury St. Edmunds Art Gallery, 2000)

Barrette, Bill, *Eva Hesse: Sculpture* (Timken Publishers Inc, 1989)

Harper, Paul, *Michael Brennand-Wood: Forever Changes* (Ruthin Craft Centre & Hare Print Press, 2012)

Johnson, Pamela, *Michael Brennand-Wood: You Are Here* (Hare Print Press, 1999)

Britton Newell, Laurie, *Out of the Ordinary: Spectacular Craft* (V&A Publishing, 2007)

Butcher, Mary, *Contemporary International Basketmaking* (Merrell Holberton in association with the Crafts Council, 1999)

David Nash at Yorkshire Sculpture Park Exhibiton Guide (Yorkshire Sculpture Park, 2010)

Dormer, Peter, *The Culture of Craft* (Manchester University Press, 2019)

Fielding, Amanda, *Gillian Lowndes* (Ruthin Craft Centre in association with York Museums Trust, 2013)

Fisch, Arline, *Textile Techniques in Metal: for Jewelers, Textile Artists and Sculptors* (Lark Books, Ashville, 1996)

Goldsworthy, Andy, *Written by Andy Goldsworthy* (Penguin Books, 1990)

Goss, Andrew, *Concrete Handbook for Artists: Technical Notes for Small-scale Objects* (Goss Design Studio, 2006)

Hogan, Joe, *Bare Branches, Blue Sky* (Wordwell Ltd for joehoganbaskets, 2010)

Johnson, Pamela, Ideas in the Making: Practice in Theory (Crafts Council, 1998)

Kessler, Robe, Stuppy, Wolfgang, *Seeds: Time Capsules of Life* (Papadakis Publisher, 2006)

Parker, Steve, *Extinction, Not the End of the World?* (Natural History Museum, 1988)

Pearson, Richard, *Driven to Extinction: The Impact of Climate Change on Biodiversity* (Natural History Museum, 2011)

Russon, Kath, *Handmade Silk Paper* (Search Press, 1999)

Scott, Jac, *Textile Perspectives in Mixed-Media Sculpture* (The Crowood Press, 2003)

Wolff, Colette, *The Art of Manipulating Fabric* (Krause Publications, 1996)

Index

Image references in italics

Acknowledgements

Image Credits

I would like to express my deepest thanks to the following people: the artists who kindly allowed me to enrich the book with their images and words; Carole Whitehill and Maisie Hulmston for being inspirational tutors whose unwavering belief in my potential gave me the confidence to become an artist; Jae Maries and Shuna Rendel for their friendship, support, encouragement and insightful suggestions; Batsford Books for offering me this opportunity, especially Lilly Phelan for giving me valuable advice and help to bring the project to conclusion; and finally my family who encouraged me to take on the project, with special thanks to Phil whose constant support, reassurance, and optimism made completion of the book possible.